T0197936

Ground Gas Handbook

Steve Wilson, Geoff Card and Sarah Haines

Whittles Publishing

Published by
Whittles Publishing,
Dunbeath,
Caithness KW6 6EY,
Scotland, UK
www.whittlespublishing.com

Distributed in North America by
CRC Press LLC,
Taylor and Francis Group,
6000 Broken Sound Parkway NW, Suite 300,
Boca Raton, FL 33487, USA

© 2009 S Wilson, G Card and S Haines
ISBN 978-1904445-68-5
USA ISBN 978-1-4398-0883-2

Typeset by Thomson Digital

Printed by InPrint, Latvia

Contents

4 Gas and vapour generation and migration   33

5 Site investigation and monitoring   56

The following figures can be found in the colour section between pages 148 and 149.

Key to colour section

Acknowledgements

The following people have provided invaluable help in producing this handbook, by providing information and photographs. We hope that their critical comments have also steered us in the right direction.

Bill Baker, Consultant

Dave Bateson, Bolton MBC

David Barry, DLB Environmental

Stephen Boult, University of Manchester and Salamander Limited

Nathan Boyd, Salamander Limited

Gill Callingham, Nottingham City Council

Michael Corban, Proctor Group Limited

Alan Crossfield, City of Lincoln Council

Paul Culleton, Environmental Protection Group

Ian Cutler, Card Geotechnics Limited

Deanne Gibbs, Card Geotechnics Limited

Ben Hill, Cooper Associates

Danny Hope, Rochdale MBC

Ray Murray, Liverpool MBC

John Naylor, Bury MBC

Roy Partington, Geosynthetics Limited

Nik Reynolds, Cooper Associates Limited

Sally Shaw, Greater Manchester Geological Unit

Andy Shuttleworth, SEL Environmental Limited

Sarah Stroud, RSK

Mark Todman, Greater Manchester Geological Unit

We would also like to thank all those local authority officers who attended the nationally held training workshops organised by the Chartered Institute of Environmental Health (CIEH) and CIRIA, whose questions and opinions helped to shape this handbook.

Foreword

This handbook must be welcomed as a significant contribution to the best practice guidance necessary to ensure safe and sustainable development of land affected by contamination, especially gas emissions from landfill sites and other sources.

It is the result of a partnership between leading specialists in the field and the Chartered Institute of Environmental Health (CIEH), which recognised the need for such guidance, and their responsibility to the local authority regulators for the provision of the training and technical guidance which government departments and their agencies are unable to provide.

Their awareness of this need arose from their experience when networking with local authority contaminated land officers (CLOs) in 2005 when delivering a programme of training courses on ground gas characterisation.

The assessment of risks from ground gas is an essential requirement of Planning Policy Statement PPS 23 and the Building Regulations Approved Document Part C, and needs to be carried out to support any proposal for development on or near land affected by gas emissions. The risk assessment needs to be submitted by the developers' consultant to the local planning authority. It must also include proposals for remediation strategies which incorporate the means to reduce the risks to an acceptable level, by preventing gas ingress and providing for its effective dispersion. Risk assessment of ground gas also plays an important role on many sites when considering whether or not a site is contaminated under Part 2A of the Environmental Protection Act 1990.

This handbook is intended to provide guidance for the CLOs, who are responsible for the assessment and regulation of all contamination-related issues in the context of both Part 2A and the planning process, and also the engineers who design the on-site gas control measures and incorporate them into the building structures.

The CIEH and the CLO community are indebted to the present authors for this important contribution to the process of effective regulation and safe development.

Dr Bill Baker

Independent consultant and Contaminated Land Advisor to the CIEH

Glossary

Active ventilation   Ventilation using fans

Advection   Transportation of contaminants by the flow of a current of water or air

Aerobic   In the presence of oxygen

Anaerobic   In the absence of oxygen

Asphyxiation   Unconsciousness or death due to lack of oxygen

Bentonite   A clay composed mainly of the clay mineral montmorillonite. It swells to many times its dry volume when in contact with water

Biodegradation   Decomposition of organic matter by micro-organisms

Borehole   A hole drilled in the ground in order to obtain samples of soil or rock. Permanent gas monitoring wells can be installed in a borehole

Borehole flow rate   Volume of air per unit time which is escaping from the borehole or standpipe (regardless of composition), usually measured in l/h

Concentration   Proportion of the total volume of void space occupied by a particular gas (denoted as % v/v). This is known as a volumetric concentration and is used throughout this handbook

Conceptual model   A theoretical representation of the ground below and around a site, including potential gas sources, migration pathways, receptors and natural barriers to gas migration

Degradable material   Any material that can biodegrade to produce ground gas

Diffusion   Movement of a fluid from an area of higher concentration to an area of lower concentration. Diffusion is a result of the kinetic properties of particles of matter. The particles will mix until they are evenly distributed

Explosion   A sudden increase in volume and release of energy in a violent manner, usually with the generation of high temperatures and the release of gases. An explosion causes pressure waves in the local medium in which it occurs

Factor of safety	A design factor, in the case of this handbook, used to provide for the possibility of gas flows greater than those assumed and for uncertainties in monitoring and modelling of gas flows in the ground and the performance of gas protection measures
Flammable	A substance capable of supporting combustion in air
Gas	One of three states of matter, characterised by very low density and viscosity (relative to liquids and solids), with complete molecular mobility and indefinite expansion to occupy with almost complete uniformity the whole of any container
Gas flow rate	Volume of gas moving through a permeable medium or issuing from a monitoring well per unit of time
Gas generation rate	Rate at which a source degrades to produce methane or carbon dioxide gas. Measured as a volume of gas produced per unit mass or volume of substrate per unit time
Gas screening value (GSV)	Gas concentration measured in a monitoring well multiplied by the measured borehole flow rate
Geomembrane	A relatively impermeable polymeric sheet used as a barrier to prevent the migration of ground gas or vapours
Geostatic stress	Static stress in the ground
Ground gas	Gas generated in the soil pore structure by processes such as biodegradation of organic material or chemical reactions, etc. Landfill gas is a specific type of ground gas
Hazard	A substance, feature or situation that has the potential to cause harm to the environment, property, humans or animals
Headspace	A term used to describe an air space in a vessel, borehole, test cell, standpipe etc., in which gas from the soil may accumulate
Hydrocarbon	A compound containing both hydrogen and carbon. In the context of this handbook it includes contaminants such as: fuels, oils, solvents, chlorinated solvents etc.
Intrinsic permeability	A measure of the ease with which fluids can flow through a porous medium. It is independent of gravity or the properties of the fluid
Landfill	Waste or other materials deposited into or onto the land

Landfill gas	Variable mixture of gases generated by decaying organic matter within a landfill site. Principal components are methane and carbon dioxide but it can contain many other trace gases and vapours
Lateral stress	Horizontal stress in the soil
Microfabric	Texture or appearance of a geological material as viewed by a petrographic microscope or by an electron microscope
Monitoring well	A well installed into the ground (commonly in a borehole but can be driven into the ground) to monitor ground gas
Oxidation	Addition of oxygen, removal of hydrogen or loss of electrons during a chemical reaction
Passive ventilation	Ventilation that relies on wind and temperature differences to create air movement
Pathway	Route by which a hazard can reach a receptor
Permeability	A measure of the ease with which a fluid can flow through a material. Normally taken to be the coefficient of permeability of water
Receptor	Environment, property, humans, animals or anything else that could be affected by a hazard
Risk	Probability that harm will occur as a result of exposure to a hazard
Soil vapour	Produced from soil as a result of volatilisation from contamination (such as petroleum hydrocarbons) that has spilt or been placed into the ground
Surface emission rate	Rate at which gas is emitted from a unit area of the ground surface, i.e. a volume of gas per unit time per unit area, usually measured in $l/m^2/h$
Trace gas	Minor constituent of ground gas
Volatilisation	Process whereby a liquid contaminant is converted into a gas or vapour
Worst credible event	An event or outcome based on the worst parameters in a data set that can physically occur in combination
Worst possible event	An event or outcome based on the worst parameters that occur in the data set irrespective of whether or not they can physically occur in combination
Zone of influence	Volume of ground surrounding a monitoring well that can be influenced by the presence of the well

CHAPTER ONE

Introduction

This handbook provides practical guidance for contaminated land officers, building control officers and others who need to undertake, manage or review ground gas assessments and design appropriate protection measures. It explains how to undertake an assessment for ground gases, including landfill gas and hydrocarbon vapours as well as gases from natural sources such as peat.

It provides a concise and practical guide of the data that is required to undertake an assessment of ground gas, including the sources and properties of key gases and vapours. It also provides guidance on site investigation and risk assessment for ground gases. Finally, it describes the various methods available to protect developments from ground gas and highlights the important points that need to be verified during their design and construction. It is intended to provide practical information about the application of various assessment and design methods. It makes extensive reference to other documents published by CIRIA and British Standards Institution (BSI) (CIRIA reports, BS 8485 and BS 10175) that provide guidance on the various methods of ground investigation and gas and vapour monitoring. The reader should be familiar with these.

Ground gas is often dismissed as a low risk due to its ubiquitous presence in small quantities below many development sites. However, this attitude is misguided. Although ground gas poses a very low risk on many sites, it is equally true that on some sites it poses a very serious risk of causing an explosion, asphyxiation or harm. Some events that have occurred due to the presence of ground gases are summarised in Section 1.1.

There are currently two recently published guidance documents in the UK relating to risk assessment and design of protection measures for ground gas. These are CIRIA Report C659/C665 (Wilson *et al.*, 2006, 2007) and the National House-Building Council (NHBC) document on the evaluation of ground gas (Boyle and Witherington, 2007). This handbook makes extensive reference to these guidance documents and the risk assessment approaches that they give. The original CIRIA Report C659 was withdrawn in 2007 due to a number of editorial issues and was re-issued as Report C665. However, the technical guidance within the document remained the same. BS 8485: 2007 makes extensive reference to the CIRIA and NHBC reports.

This handbook considers ground gas from a development perspective or situations where a site is being assessed under Part 2A of the Environmental Protection Act 1990. The handbook is not intended as guidance on managing gas in currently licensed and operational landfill sites, as these are covered by specific waste management legislation. There is also a wealth of information and guidance published by the Environment Agency that relates to operational sites and their specific requirements that provides useful technical information (e.g. Environment Agency, 2004).

1.1 Landfill gas incidents

Although explosions caused by ground gas are by no means a common occurrence, incidents where death or serious injury has resulted from suspected ground gas explosions have been documented. A selection of these incidents is listed below. All are associated with migration of landfill gas from landfill sites. The explosions were caused by the methane within the landfill gas. Detailed information about the background to the incidents can be obtained from the specific references.

- 2000. A home exploded in Rochester, Michigan when landfill gas was ignited by a pilot light (*Chartwell's Weekly News Update*, 2002)
- 1999. An eight-year-old girl was burned on her arms and legs by an explosion from landfill gas when playing in a playground in Atlanta, Georgia. The area was reported to have been used as an illegal dumping ground many years previously (*Atlanta Journal–Constitution*, 1999)
- 1994. While playing soccer in a park built over an old landfill in Charlotte, North Carolina, a woman was seriously burned by a landfill gas explosion (*Charlotte Observer*, 1994)
- 1987. A gas explosion occurred in a house next to a landfill site at Stone, Kent, UK (Health and Safety Executive, 2003)
- Circa 1987. A house 50 m from a waste-filled limestone quarry was seriously damaged by a landfill gas explosion. The gas had migrated along natural fissures in the limestone underneath the house (www.landfill-gas.com)
- 1987. It is suspected that off-site landfill gas migration caused a house to explode in Pittsburgh, Pennsylvania (United States Environmental Protection Agency (USEPA), 1991)
- 1986. Explosion in a house caused by landfill gas migration from nearby landfill site in Loscoe, Derbyshire, UK (Williams and Aitkenhead, 1989)
- 1984. Landfill gas migrated to and destroyed a house near a landfill in Akron, Ohio. Ten houses were temporarily evacuated (USEPA, 1991; New York Times, 1984)

- 1983. An explosion destroyed a residence across the street from a land-fill in Cincinnati, Ohio. Minor injuries were reported (USEPA, 1991)
- 1975. In Sheridan, Colorado, landfill gas accumulated in a storm drain-pipe that ran through a landfill. An explosion occurred when several children playing in the pipe lit a candle, resulting in serious injury to all the children (United States Army Corps of Engineers (USACE), 1984)
- 1969. Landfill gas migrated from an adjacent landfill (about 30 m away) into the basement of an armoury in Winston-Salem, North Carolina. A lit cigarette caused the gas to explode, killing three men and seriously injuring 12 others (USACE, 1984)
- 1965. A boy playing in a cave he had dug in his backyard in California was burned in a flash fire when he attempted to light a candle. The gas was attributed to the nearby Palos Verdes landfill (Hickman, 2001)
- 1947. A fire in a sewer manhole was caused by landfill gas migrating from the overlying refuse (Bailer, 1947)

In addition to incidents where people have been injured or killed there have been numerous occurrences in the UK where landfill gas has migrated from landfill sites and been detected within buildings on adjacent sites (Health and Safety Executive, 2003). A selection of incidents is listed below:

- 2005. Audenshaw, Manchester, UK, landfill gas migrated from a former clay pit which was being used as a landfill site. Gas alarms were fitted to four properties after gas was detected during a routine inspection. A vent trench was also installed (House of Commons debate, 1 November 2005)
- 1992. Inverness, Scotland, landfill gas was detected below an industrial estate adjacent to a landfill site. Monitoring was required
- 1992. Airdrie, Scotland, landfill gas was detected in buildings adjacent to a landfill site. Continuous monitoring and ventilation were required
- 1990. Barnsley, South Yorkshire, UK, landfill gas was detected in a factory adjacent to a landfill site
- 1990. Thurmaston, Leicestershire, UK, landfill gas entered houses built on a former landfill site. Venting and monitoring were required
- 1988. Appleby Bridge, Lancashire, UK, partial blockage of a gas venting trench was thought to have allowed gas to migrate into an office building some 50 m away. It resulted in an explosion which caused structural damage
- Date not known. Crowborough, East Sussex, UK, landfill gas was detected below an industrial estate adjacent to a landfill site. Venting and monitoring was required

1.2 Are other sources of ground gas a risk?

Ground gas from other sources is often treated less seriously than when a landfill site is the source. There are recorded incidents where other sources of ground gas have been the cause of death or injury. The most common incidents are explosion or asphyxiation caused by methane and carbon dioxide in coal mines. However, mine gases can migrate to the surface through adits, shafts and fissures and cause serious problems in buildings. Geological events such as volcanic activity, although less frequent, can also emit large volumes of gas very rapidly. Some incidents are listed below:

- 2000. A home owner and grazing farm animals were asphyxiated by a carbon dioxide gas cloud arising from natural volcanic activity near Cava del Selci, Italy
- 1999. A construction worker was killed in a trench near Barnsley, South Yorkshire, UK, when carbon dioxide collected in the base of the trench. The trench had been constructed through colliery spoil in an area of historic mining (*New Civil Engineer*, 1999)
- 1995. A man died when carbon dioxide seeped into a factory from a mine entry in Widdrington, Northumberland, UK (Department of the Environment, 1996)
- 1990s. Degradation of a clayboard void former below a basement floor slab generated methane that collected in a basement and caused an explosion (Hooker and Bannon, 1993)
- 1988. Elevated concentrations of methane accumulated in several houses in Arkwright Town, Derbyshire, UK as a result of migration from mineworkings. Evacuation was required until remedial measures were installed. Eventually, the whole village was relocated (Department of the Environment, 1996)
- 1987. Methane migrated into a house via a backfilled mineshaft in Haydock, Lancashire, UK (Barry and Raybould, 1988)
- 1985. Methane from deep geological deposits seeped into a department store in Fairfax, California and caused an explosion that injured 23 people (Council of the City of Los Angeles, 2004)

Historically there have also been incidents of methane migrating from coal mines and causing explosions in coal mining areas in the UK.

As far as the authors are aware there have not been any explosions or other incidents in the UK caused by ground gas migration into above-ground buildings from natural sources such as alluvium, chalk, limestone and peat. There have not been any recorded incidents where either methane or carbon dioxide have migrated from undisturbed coal measures into buildings.

An explosion occurred in a pumping station at Abbeystead, Lancashire, UK in 1984 (Hooker and Bannon, 1993). This was caused by methane accumulating in a tunnel constructed in coal measures strata. The source of the methane was not proved but was thought to be a deep source below the tunnel. The methane migrated to shallower depths in solution in groundwater and accumulated in fissures in the rock where the tunnel was constructed. The methane entered the tunnel as free gas and dissolved in groundwater.

1.3 Radon

Radon is a naturally occurring radioactive gas that is odourless and colourless. It is formed where uranium and radium are present in the soil (at very small concentrations) and can migrate vertically into any buildings built over the source. It poses a chronic rather than acute risk because if it accumulates in a building at unacceptably high concentrations it will increase the risk of lung cancer.

Dealing with radon is quite straightforward as risk assessment is not usually required. Guidance on the need for protection is given in Building Research Establishment Report BR 211 (BRE, 1999). This includes maps that show where radon protection is and is not required (full or basic protection), or where a further assessment is required (this is carried out by the British Geological Survey). It also provides guidance on the design requirements for basic and full radon protection measures.

Radon is not discussed further in this book, although many of the comments in Chapter 7 regarding good practice for venting and membrane installation also apply to radon protection measures.

SUMMARY

This handbook provides a concise reference for anyone involved with ground gas assessments and design protection measures. It explains how to undertake an assessment for ground gases, including landfill gas and hydrocarbon vapours as well as gases from natural sources such as peat.

Ground gas poses a very low risk on many sites, but it is equally true that on some sites it poses a very serious risk of causing an explosion, asphyxiation or harm. There are several instances from around the world where landfill gas, and mine gas in particular, have caused explosions or asphyxiation. The most commonly known events in the UK were at Loscoe in Derbyshire and the Abbeystead pumping station.

There have not been any explosions or other incidents reported in the UK that have been caused by ground gas migration into buildings from natural sources such as alluvium, chalk, limestone or peat. There have not been any recorded incidents where either methane or carbon dioxide have migrated from undisturbed coal measures.

Dealing with radon is quite straightforward and guidance on the need for protection is given in Building Research Establishment Report BR 211 (BRE, 1999).

CHAPTER TWO

Regulations and policy guidance

UK Government policy for managing contaminated land is that it should be achieved primarily through the planning and building control process when sites are redeveloped. There are three main documents that define the government's approach to managing risks associated with ground gas when considering redevelopment of sites. These are:

- Planning Policy Statement 23 (PPS 23)
- Environment Agency policy on development within 250 m of a landfill
- Building Regulations Approved Document Part C

Where sites are not likely to be dealt with as part of redevelopment in a reasonable timescale, local authorities can use the powers granted to them under Part 2A of the Environmental Protection Act 1990 to assess risks posed by sites and ensure that they are remediated where necessary.

The design of gas protection systems may also require consideration of legislation relating to health and safety and waste management. There is a wealth of technical guidance specifically about landfill gas that is available from the Environment Agency, primarily intended for operational and currently licensed landfill sites. However, much of the technical information can also be applied to the assessment of landfill or other ground gas below development sites (Environment Agency, 2003, 2004a-d). There is very little regulation, relating to either the maintenance of gas protection systems or on climate change, for the design of systems that vent methane and carbon dioxide (greenhouse gases) to the atmosphere.

2.1 Planning Policy Statement 23

PPS 23, Planning and Pollution Control (Department for Environment, Food and Rural Affairs (DEFRA)) sets out UK Government policy for planning in relation to land affected by contamination. Land contamination is referred to in Annex 2 of that document. The key points that are relevant to ground gas issues are:

- The presence of contamination, whether man-made or natural, is a material consideration in the planning process
- Local development plans should identify constraints to site use at an early stage to try and match the type of development to the constraints (e.g. do not zone a landfill site for housing)
- Wherever possible developers should arrange pre-application meetings with the planners, environmental health department, building control and the Environment Agency to agree the issues surrounding contamination and agree a scope of investigation and risk assessment. This is especially important in relation to ground gas where some of the guidance and regulations can have widely varying interpretations placed on them. Many local authorities have produced their own supplementary planning guidance documents relating to contaminated land which may be useful at this stage
- The developer (assisted by advisors) is responsible for satisfying the local authority that any unacceptable risk will be managed satisfactorily way. They must provide all necessary information. Developers and their consultants should realise that local authorities usually have little time to review a site that the consultants themselves have intimate knowledge of. It is unwise to assume that the local authority has access to comprehensive information about every site within their area. The more comprehensive the data and the better it is presented, explained and summarised, the easier (and more importantly quicker) it is for the regulators to make decisions. It should also be remembered that if a regulator asks a question it must be answered and all recommendations must be justified with sound facts, reason and supporting references where applicable. Copies of the base data from which assumptions have been made or conclusions drawn should be provided
- All site investigations should comply with British Standard BS 10175: 2001 (Investigation of potentially contaminated sites: a code of practice). The PPS states that site investigations should be carried out under the direction of a suitably qualified and competent person who is a chartered professional. Gas risk assessment may be complex and practitioners require interdisciplinary understanding of a number of fields including geology, physics and chemistry. Therefore, in line with the PPS and guidance produced by the Association of Geotechnical and Geo-environmental Specialists (AGS) (AGS, 1998) for site investigations and geotechnical assessment, it is recommended that ground gas investigations, risk assessments and design of protection measures should be approved

by a chartered professional who specialises in geo-environmental issues and has appropriate experience of contaminated land and ground gas risk assessments

- A desk study and site investigation report should be submitted with a planning application
- The local authority needs to consider any adverse impacts on the surrounding environment that may occur during, or as a result of, remediation (e.g. gas or vapour emissions from excavating in a landfill or from venting points in gas protection systems)
- Regulators need to apply conditions to planning permission to ensure that all risks are safely managed. For gas this may mean conditions are applied that require off-site migration and migration into buildings to be controlled
- Conditions may include requirements to provide adequate provision for long-term maintenance and transfer of responsibility for maintenance if the ownership of a site were to change in the future
- Planning conditions may include a 'pre-commencement' clause. If development starts prior to the discharge of the conditions, the development becomes unauthorised and may be subject to enforcement or stop action
- Extreme caution is recommended in granting outline planning permission where land contamination is an issue, unless it is fully demonstrated in detail how remediation will be undertaken
- Planning permission can be refused if the land contamination issues are not fully addressed or if sufficient information on the ground conditions has not been provided
- Planning conditions can (and should) include requirements for verification of any remediation works or installation of gas protection systems. These usually include a pre-occupation clause
- Planning obligations can be used to secure the off-site treatment works needed to protect a development (e.g. an off-site gas migration barrier) or to secure funds to ensure the maintenance of protection measures

2.2 Environment Agency policy on development within 250 m of a landfill

The Environment Agency policy (Environment Agency, 2003c) states:

> For applications for building development within 250 metres of a landfill site authorised by the Agency, the Agency recommends that the LPA (Local Planning Authority) obtains a risk-based assessment of the impact of any emissions from the landfill on the development site. These proposals should

include detailed risk management actions to deal with any risks identified. Failing this the LPA should refuse the application.

Authorised sites are those which have either:

- a permit issued under the Pollution Prevention and Control Act 1999, or
- a waste management license issued under Part II of the Environmental Protection Act 1990

In the case of sites where no license or permit exists the Agency will not make any recommendations but will provide any relevant readily available information it holds to help the LPA make its decision.

It should be noted that this is only applicable to landfill sites and not other sources of ground gas.

The document notes that carrying out a comprehensive risk assessment is time consuming and expensive. A tiered approach should be taken. On many sites the initial risk screening will be sufficient to characterise the risk. It is, however, essential that a full site investigation and an appropriate level of risk assessment is carried out and that it is demonstrated that any risk management measures that are proposed are sufficient to reduce the risk of harm to the development and its occupants to an acceptable level for as long as the hazard exists.

2.3 Building Regulations Approved Document C (revised edition)

The Building Regulations Approved Document C was revised in 2004. One of the key changes was that the old 'trigger levels' for methane and carbon dioxide were removed (1% and 1.5% v/v, respectively). The guidance now requires a risk-based approach to the assessment of ground gas. The key points are listed below:

- The requirements in relation to contamination apply not only to buildings but also the surrounding land. Therefore there is a need for approved building control inspectors to consider the effects of gas on gardens (e.g. sheds, vegetation, the effect of digging ponds etc.)
- The approved document relies heavily on other sources of guidance on ground gas
- It identifies the need to measure not only gas concentrations in the ground but also to collect information to assess: the volume of gas generating material, the rate of generation, the gas movement in the ground and surface emissions
- Remedial measures to buildings usually include gas resistant membranes over the footprint of the building and a vented underfloor void (normally passively vented). For all but the lowest risk sites both methods of protection are usually required as a minimum

2.4 Part 2A of the Environmental Protection Act

Part 2A of the Environmental Protection Act 1990 was introduced by the
Environment Act 1995. It came into force on 1 April 2000. Contaminated land
is defined in Part 2A as follows:

> any land which appears to the local authority in whose area it is situated to be in
> such a condition, by reason of substances in, on, or under the land that:
>
> (a) significant harm is being caused or there is a significant possibility of
>   such harm being caused; or
>
> (b) pollution of controlled waters is being, or is likely to be caused.

Local authorities have a duty under the Act to identify land within their area
that is likely to fall under this definition. If following investigation, the land
is formally determined to be contaminated, the local authority must ensure
that it is remediated to an acceptable standard. Formal determination that land
is contaminated is based on identifying significant pollutant linkages between
sources of contamination and receptors via an existing migration pathway.
It should be noted that Part 2A only considers current land use and existing
migration pathways.

Significant receptors for gas are not only humans or buildings but can also
include building services, crops, farm animals or sites of ecological importance.
Specific receptors for a site should be identified.

Significant harm can include: death, serious injury, disease, reduction of crop
yields or structural damage to buildings. Many risk assessments only consider
death or injury to humans. It is clear that there are other factors that need to be
considered in any risk assessment to formally determine whether or not a site
is contaminated due to landfill or other ground gas.

Government policy in the UK is to remediate land via voluntary means
wherever possible. The preferred route is by encouraging redevelopment of
contaminated sites and using the planning and building control systems to
obtain satisfactory remediation (Scottish Environmental Protection Agency
(SEPA), 2007). Using the powers of Part 2A is seen as a method of last resort.

When assessing risk in relation to Part 2A there is a very clear need for
comprehensive information and robust data in order to make a reasonable assessment of risks. This is far more important for this situation than
with development sites where a lack of data can usually be dealt with by
a conservative approach to the design of remedial measures. Being overly
conservative in a Part 2A assessment of risk can be just as bad as being optimistic. For example, if the over-conservatism results in properties that are not
at risk being blighted, or an unnecessary cut-off barrier is installed at great

cost. Although many of the principles discussed in this handbook can be applied to Part 2A assessments they should be undertaken with care. In particular, the advice regarding periods of gas monitoring in Section 5.2 will not be applicable, much longer periods of monitoring are likely to be required. The use of the continuous monitoring equipment that is now available will be invaluable in providing data for Part 2A assessments (see Chapter 5).

2.5 Legal requirements for maintenance

There are no specific legal requirements to maintain gas protection measures once they have been installed, although it will often fall under the requirements of general health and safety legislation. The Planning and Building Control legislation is only enforced up to the point of completion of construction. One exception is stadia, where the regulatory regime relating to stadium safety certificates can be used to ensure the maintenance of any gas protection systems. Requirements for long-term maintenance are therefore generally difficult to enforce. Only in a few rare cases around the UK are gas protection systems regularly inspected by the local authority.

Although maintenance cannot generally be enforced by a local authority, it can be encouraged by regular (annual or bi-annual) reminders to the site owners or operators of the implications of failure to maintain the gas protection. This could result in a site being investigated under Part 2A, or litigation if poor maintenance leads to damage to person or property.

If an incident such as an explosion did occur because of lack of maintenance of gas protection measures in a commercial or similar building, then it is likely that the owner and/or operator would be prosecuted under Health and Safety legislation. A similar scenario would be where the owner of a gas barrier did not maintain it and this caused gas migration to another site and resulted in an incident such as an explosion.

2.6 Climate change

Methane and, to a lesser degree, carbon dioxide are greenhouse gases that are considered to contribute to climate change. For this reason the emission of such gases from licensed landfills is required by legislation to be minimised (for example by re-using the gas to generate electricity). However, for most development sites the gas generation is usually too low to enable electricity generation or even flaring. If gas can be extracted for re-use then the site is probably not suitable for redevelopment due to the gas risk.

However, where appropriate, it is possible to promote the oxidation of methane to carbon dioxide in venting systems. If this approach is considered then careful assessment of the likely benefits as against other environmental costs needs to be undertaken. For example using a material to oxidise the methane may reduce the flow rates and require some form of fan system that uses electricity, possibly negating any beneficial effect on the emissions.

2.7 Waste management

The present authors are aware of cases where gassing material from excavations on sites has been placed below buildings, on the basis that the gas protection is already required and that this will not increase the risk. However, moving waste in such a manner can change the gassing regime and the generation rate may change as the oxygen content, moisture content etc. changes. The volume of gassing material will also be increased. Therefore placing additional gassing material below buildings is considered unacceptable and should not be approved.

In addition to being bad practice, gassing material will be considered a waste and excavating or moving such material on a site will require the approval of the Environment Agency and will fall under the waste management legislation. It is possible that a waste management site license will be required that will place very onerous long-term monitoring and management requirements on the owners of the site.

SUMMARY: Regulations and policy guidance

UK Government policy in relation to managing contaminated land is that it should be achieved primarily through the planning and building control process when sites are redeveloped. Guidance on good practice for developers, their consultants and local authorities when submitting planning applications for contaminated sites is given in PPS 23. There are also requirements for dealing with contamination, including ground gas, in the Building Regulations. The Environment Agency provide some generic advice on assessing risks for developments that are located within 250 m of a landfill site.

For sites that fall outside the development and planning context, local authorities have the power to assess sites and enforce remediation where necessary, under Part 2A of the Environmental Protection Act 1990. UK Government policy is to remediate land via voluntary means wherever possible by encouraging redevelopment of contaminated sites and using the planning and building control systems to obtain satisfactory remediation. Using the powers of Part 2A is seen as a method of last resort. When assessing risk in relation to Part 2A there is a very clear need for comprehensive information and robust data in order to make a reasonable assessment of risks. This is far more important for this situation than with development sites where a lack of data can usually be dealt with by a conservative approach to the design of remedial measures.

PPS 23 states that the presence of contamination (which includes ground gas), whether man-made or natural, is a material consideration in the planning process. It includes a list of good practice guidance about what information should be provided to a local authority to support a planning application. Much of this concerns communication between the developer and the local authority to avoid delays in the approval process. Regulators need to apply conditions to planning permission to ensure that all risks are safely managed. For gas this may mean conditions are applied that require off-site migration and migration into buildings to be controlled. Planning permission can be refused if the land contamination issues are not fully addressed or sufficient information on ground conditions has not been provided.

The Building Regulations Approved Document C now requires a risk-based approach to the assessment of ground gas. It relies heavily on other sources of guidance on ground gas, especially the suite of reports on ground gas published by CIRIA.

There are no specific legal requirements to maintain gas protection measures once they have been installed although it will often fall under the requirements of general health and safety legislation. If an incident, such as an explosion, did occur because of lack of maintenance of gas protection measures in a commercial or similar building, then it is likely that the owner and/or operator would be prosecuted under the Health and Safety legislation.

Methane and, to a lesser degree, carbon dioxide are greenhouse gases that are considered to contribute to climate change. Where appropriate, it is possible to promote oxidation of methane to carbon dioxide in venting systems.

Placing additional gassing material below buildings is considered unacceptable and should not be approved. In addition to being bad practice, gassing material will be considered as a waste and excavating and moving such material on a site will require the approval of the Environment Agency and will fall under the waste management legislation.

CHAPTER THREE

Sources and properties of ground gases and vapours

3.1 Sources of ground gas

In addition to landfill sites there are several other man-made or natural sources which contain organic materials that can be degraded by bacteria to produce gas. Gas can also be trapped in material such as coal and be released when excavations for mines take place (or in peat and released when activities such as piling occur). Each source has different generation patterns in terms of volume and duration and as a result the level of risk associated with differing sources varies.

The various sources and typical level of risk associated with varying sources is summarised in Table 3.1 which is based on the guidance provided in CIRIA Report 152 (O'Riordan and Milloy, 1995). Any such assessment of the source can only provide an initial indication of the likely risks posed by the various sources. Because of the variability in the nature of sources, particularly any made ground or landfill, an initial assessment made using Table 3.1 should always be confirmed by a site investigation.

Generation rates are only provided in descriptive terms as it is virtually impossible to assign typical values to the different sources, due to the number of variables involved. The generation rates in Table 3.1 are provided in order to allow a comparison of the relative risks posed by the various sources. The main factors that affect landfill gas generation and lateral migration are discussed in more detail in Section 4.1.

At the worst extreme a very high generation rate can be taken as 10 m^3 per tonne of waste per year or higher, which is the typical gas generation rate of a new landfill site (Nastev, 1998 (see Chapter 4)). At the opposite end of the spectrum very low generation rates mean negligible volumes of gas are being produced that are at least 10 or 100 times lower than the peak rate from a new landfill site. Carbon dioxide can be produced by natural soils at very low rates that results in elevated concentrations within monitoring wells. Carbon dioxide can be present in wells that are installed in soils or rocks that include carbonate content such as chalk and limestone (or clay with chalk gravel).

There are landfill sites that have been filled predominantly with industrial waste. They are difficult to assess without a detailed knowledge of what materials have been placed in them, so they are not included in Table 3.1. They can be compared to domestic landfill sites in terms of gas generation. However, if they have been filled with predominantly chemical waste then the gas generation rates are often very low, because the limited degradable content and/or the chemical make up retards or prevents the degradation process occurring.

Sources of vapours are numerous and include any situation where there has been spillage or storage of volatile materials. This includes petrol filling stations, chemical industry sites and similar locations. However, historically many places such as bakeries and small factories had their own oil tanks or tanks for storing degreasing chemicals etc., so the presence of hydrocarbon vapours is very widespread (in the context of this handbook, 'hydrocarbon' relates to any compound containing both carbon and hydrogen, including solvents, chlorinated solvents etc.). Vapours occur due to volatilisation of a solid or liquid rather than biodegradation so they do not have a generation rate as such. They volatilise at very low rates compared to the generation of landfill gas in a domestic landfill site, but need careful consideration because adverse health effects occur at relatively low concentrations. Some volatile compounds can also easily migrate via groundwater due to their relatively high solubility.

It is possible for large volumes of gas to be generated from sources that have a low organic content but where the total volume of gassing material is very large. There has been an example where fill material with a total organic carbon content of less than 5% generated large volumes of gas because the organic carbon had been finely shredded and distributed through the fill. The volume of the fill was some 2 million m^3.

Table 3.1 *Examples of ground gas sources*

Source	Generation potential	Level of risk for on site development	Risk of lateral migration
Natural carbonate soil and strata e.g. chalk and limestone or soils with chalk gravel	Very low	Negligible	Negligible
Natural soil strata with a low degradable organic content e.g. alluvium	Very low	Very low	Negligible
Infilled pond less than 15 m diameter, infilled before 1940	Very low	Very low	Negligible

Source	Generation potential	Level of risk for on site development	Risk of lateral migration
Natural soil strata with a high degradable organic content e.g. peat (gas in peat is historically generated and is trapped or adsorbed in the soil so the actual current generation rate is very low)	Very low/low (see note to left)	Low	Negligible
Made ground with low degradable organic content (e.g. up to 5% organic material such as pieces of wood, sheets of paper, rags, etc. with a high proportion of ash. No food or other easily degradable waste)	Very low	Very low	Negligible
Made ground with high degradable organic content up to 15% (e.g. dock silt. No food or other easily degradable waste)	Low	Low/moderate	Negligible
Foundry sand (includes phenolic binders, rags and wood that decay, albeit at low rates)	Low	Low/moderate	Very low
Sewage sludge, cess pits	Moderate	Moderate	Very low
Landfill 1945 to mid 1960s inert	Low	Low/moderate	Low
Flooded mineworkings (gas is liberated from coal when mineworkings are excavated, this continues for up to 50 years)	Low/moderate	Low/moderate	Variable: depends on extent of workings, geology and hydrogeology
Landfill 1945 to mid 1960s	Low/moderate	Low/moderate	Low/moderate: depends on geology
Shallow mineworkings or shaft (not flooded) (gas in coal is historically generated and is trapped or adsorbed so the actual current generation rate is very low. However, it accumulates in workings and large volumes can be emitted very quickly)	Very low (see note to left)	Moderate/high	Variable: depends on extent of workings, geology and hydrogeology
Landfill mid 1960s to early 1990s Municipal landfill sites	Moderate to very high	Moderate to very high	Moderate to very high

Source	Generation potential	Level of risk for on site development	Risk of lateral migration
Inert landfill sites (note that lack of regulation during this period means that most sites are never entirely inert, they can include timber, plasterboard and even domestic refuse). Care is needed when assessing such sites	Low	Low/moderate	Low
Landfill early 1990s onwards			
Municipal landfill sites	High to very high	Moderate to very high	Low (assuming site has engineered containment systems)
Inert landfill sites	Low	Low	Low

[1] Inert means an absence of material that can cause generation of significant volumes of gas (i.e. typically less than 5% if large pieces of wood or trees and less than 1% if finely disseminated, e.g. sawdust)

3.2 History of landfilling

In order to understand why older landfill sites produce less landfill gas, a look at the history of landfilling in the UK is instructive. The history of landfilling provided below is not a complete discussion, particularly in relation to waste management legislation. It aims to highlight the main influences on the type and nature of waste and how the engineering of landfill sites has changed throughout history.

3.2.1 Early history

Ever since mankind started to live in small farming communities and stayed in one place there has been a need to dispose of waste. Simply burying it has always been one of the main methods of disposal. *Midden* is an old English word for a household rubbish dump. They were places where food remains, such as shellfish and animal bones, ash and charcoal from fires, and broken or worn out tools were dumped or buried. The middens were small and usually close to the house. However, the volume of waste dumped in the ground was low, as re-use and recycling was commonplace and an economic necessity. Tools and clothing were repaired, vegetable waste was used to feed livestock and manure was used to fertilise the fields. It is interesting that we are now trying manage our waste the way our ancestors did over 2000 years ago!

3.2.2 Medieval Period

By the Medieval Period rubbish was being burnt in open fires and was also thrown out into the streets in developing towns and cities. This led to the streets being covered in decomposing rubbish, as well as the sewage that was also disposed there. Towns also used free roaming pigs to eat the rubbish that was thrown out. Rats were common and led to the Black Death of 1348–1349. By the middle of the 14th century people known as *rakers* were employed in London to rake up and cart the rubbish away once a week, for disposal either in pits outside the city limits or to be taken elsewhere by boat. These were the first ancestors of modern municipal landfill sites, although the nature of the waste was very different. There would have been no distinction between municipal and industrial waste.

In 1407 it was ruled that household rubbish in London had to be kept inside until the carts arrived to take it away. This first piece of waste management legislation was largely ignored, even though in 1408 Henry IV instructed that forfeits be paid by those who ignored the law. Laws were also passed requiring people who brought goods or produce into some cities to carry solid wastes away with them for disposal in the countryside.

3.3.3 The Industrial Revolution

The Industrial Revolution led to machinery being able to produce larger quantities of products more quickly and easily than just manpower. At the same time there was a shift of population from the countryside to towns and cities and an increase in consumption. The combined effect was an increase in the volumes of household waste and the generation of distinct industrial waste.

Household rubbish was still discarded in the streets, but the traditional method of using pigs to eat much of the waste was not possible in the overcrowded towns and cities. Many people survived by scavenging through the waste and removing anything of use that could be sold. For example dog dung was used by tanners to treat leather and so was collected. Dustmen collected the ash from the hundreds of coal fires. This was taken to yards where it was sieved and the coarse fraction of the dust (the cinders known as breeze) was used in brickmaking. Animals such as pigs and chickens also rooted around the dust yards for food scraps in the dust. Any other valuable items were taken and sold (boots, kettles, rags etc.). The material that was finally buried would have been mainly inert.

There was generally no organised means of household waste disposal, although in some towns (e.g. Southampton) scavengers were appointed to keep the streets clear of rubbish. They paid a yearly fee to be allowed to collect the rubbish and householders paid scavenge money. The waste was sorted and any material that could not be recycled was then left in piles on marshland and used as fertiliser.

3.2.4 Beginning of waste management legislation

The Public Health Act 1848 was the first piece of legislation that specified how waste should be stored and removed. The Act specified that household waste should be placed in midden heaps (unlined holes in the ground) next to houses, and when they were full the waste was dug out and taken away. It also provided for a Central Board of Health to be established with powers to supervise street cleaning and refuse collection amongst other things.

The fourth cholera pandemic (1863–1875) claimed many lives in London and afterwards the Public Health Act 1875 was passed. One aspect of this was that it placed a duty on local authorities to arrange the collection and disposal of waste each week or face penalties if they failed to do so. Each resident was required to place their rubbish in a moveable receptacle: the first dustbins. The subsequent Public Health Act of 1891 required local authorities to provide a sufficient number of scavengers to ensure the streets were kept clean in their district and that household waste was collected and removed.

By the end of the 19th century household waste was collected regularly from ash bins. As with the earlier scavengers and dustmen a large proportion of the waste was sorted and recycled and the coarser cinder particles from the ash separated for use in brickmaking. The end of this century also saw the first energy from waste schemes with the construction of the first 'destructor' in Nottingham. This was an incineration plant that burnt waste as part of a mixed fuel, thus producing steam to drive electricity generating turbines. Over the last 25 years of the century 250 destructors were constructed in the UK. They were used to generate electricity and power sewage pumping stations.

As today, they were opposed on the grounds of unacceptable emissions. In those days ash and charred paper were falling onto the surrounding areas and the use of destructors gradually fell into decline in the first part of the 20th century. Recycling was still important and at some destructors tin cans and similar items were picked from the refuse before it was incinerated. In Sheffield the ash and cinder from the incinerator was taken directly to the public baths and re-burnt to heat the water (the household fires were not very efficient and the ash still had a high calorific value and could be burnt in boilers). At that time the bulk of waste disposed by households would have been ash and other inert materials (about 80%).

3.2.5 Early 20th century

In 1907 an amendment to the Public Health Act 1875 extended the requirements for local authorities to collect trade refuse and also allowed them to charge for its removal. At this time the nature of waste began to change and the use of disposable items and packaging started to develop with such things as disposable paper cups, paper handkerchiefs and razor blades. In the 1920s bins were beginning to be issued to some households in parts of the UK.

The 1936 Public Health Act was the first legislation to set out rules for landfill operation, although they were not strictly enforced and most landfills were unlined. The Act also allowed for the prosecution of people for illegal dumping of waste. At that time a large proportion of municipal waste was still ash from fires and people were actually encouraged to burn their own rubbish. Much industrial waste would have been placed in private landfill sites within factories. Most textiles, glass and metals were recycled via the reclamation schemes of rag and bone men. During the Second World War there was a great deal of recycling and composting of waste as part of the war effort (i.e. because of economic necessity rather than environmental concerns), and in some cities (e.g. Sheffield) inspectors were employed to check dustbins and ensure that any recyclable material was not thrown away.

3.2.6 After the Second World War

After the war the nature of waste began to change as society became more consumer oriented. There was a gradual increase in packaging and products designed to be thrown away when used, so the amount of waste began to increase. The use of aluminium cans grew together with the use of plastics. Manufacturing also expanded, there was less emphasis on recycling. This all added to the amount of waste being generated. For the next 40 years placing waste in unlined landfills would be the most common method of waste disposal with little thought given to managing leachate and landfill gas. Landfill sites were uncontrolled and sites were generally chosen wherever there was a convenient hole in the ground with little regard to environmental impacts. However, much of the rubbish was burnt on a daily basis at the landfill sites so the resulting material in the ground was predominantly ash and cinders.

In 1947 the Town and Country Planning Act was passed. This required all new landfill sites to obtain planning permission. The new legislation did not apply to existing sites and there was still no consideration of managing leachate or gas emissions from sites and no effort to control the types of waste accepted.

In response to air pollution in major cities the Clean Air Act 1956 was passed. As a result, the number of household open coal fires began to decline, to be replaced by gas and electric fires. Thus the burning of rubbish reduced and the volume disposed of into bins increased and changed from being mostly ash to containing an increasing proportion of degradable material such as paper and food. This resulted in the ash content reducing to about 10% by the late 1960s.

3.2.7 1970s

By the 1960s increasing amounts of industrial and chemical waste were being deposited in some landfills, and after one incident, when drums of cyanide were illegally dumped near a children's playground, the Deposit of Poisonous Waste Act 1972 was passed to try and control such activities. This introduced for the first time a notification system to control and record the volumes, movement and disposal of poisonous, noxious or polluting waste to prevent pollution of water and land.

In the 1970s waste began to be compacted using steel wheeled compactors. This reduced the void ratio and increased the density. Modern versions tend to break up the waste and make it more homogeneous, which may speed up biodegradation.

3.2.8 Control of Pollution Act 1974

Just two years after the Deposit of Poisonous Waste Act was passed it was replaced by the Control of Pollution Act 1974. Earlier legislation on waste disposal had been led by a public health perspective with no concern for the wider environmental issues related to leachate and landfill gas. There was no classification of waste and no control over what types of waste were deposited in landfill sites. There are numerous stories of dead cows, chemical drums etc. being deposited in landfill sites. Sites were usually capped off with a layer of soil to cover the waste and without any thought of controlling gas emissions or rainfall infiltration.

The Control of Pollution Act 1974 was the first major legislation to control the operation of landfill sites, with an emphasis on environmental protection. It led to the creation of waste collection authorities and introduced licensing of waste disposal facilities. The Act lead to the introduction of waste classification and the first controls on what type of waste landfill sites could accept. The main classifications are shown in Figure 3.1.

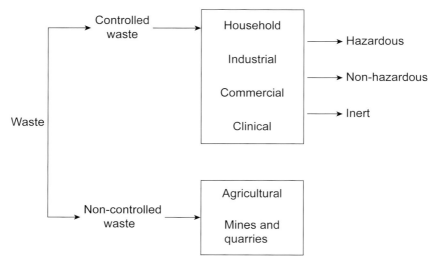

Figure 3.1 *Waste classification*

The Act also required consideration of the wider environmental effects of waste disposal due to leachate generation. However, there were some serious faults in the Act including the following:

- The regulatory authorities had limited powers to refuse license applications or control their transfer between parties
- Licenses could be surrendered at will
- Little consideration was given to landfill gas

At that time dilute and disperse was still an accepted practice, most landfills were not lined and large volumes of degradable waste was placed in unlined pits in permeable soils (i.e. the worst conditions for landfill gas and leachate generation and migration). Co-disposal of hazardous waste with other types was also an accepted practice. Thus some domestic refuse sites from that period can include a wide range of hazardous materials.

The 1970s and 1980s saw a huge rise in disposability with many items now being designed to be disposed of rather than repaired, for example shoes, furniture and clothes. The consumption of fast food and packaging also increased and the density of waste decreased, all leading to increased volumes of degradable and toxic waste. Other items such as cookers and fridges were being thrown away in increasing numbers. The use of batteries and electrical equipment was also rising, with associated toxic compounds being present in waste.

The increase in waste and the change in its composition was reflected in the need to increase the size of dustbins because the old metal bins were no longer large enough. Most town and cites started using larger wheelie bins.

The Control of Pollution (Special Waste) Regulations 1980 were implemented under the Control of Pollution Act 1974. Hazardous waste was defined through toxicity criteria and the concentration of chemicals or substances. It retained the need (from the Deposit of Poisonous Waste Act 1972) to pre-notify the waste regulation authorities before moving special waste to pre-determined disposal sites.

In 1989 the first edition of Waste Management Paper 27 was published by the Department of the Environment (it was updated in 1991). This provided the first guidance on managing landfill gas emissions from landfill sites.

3.2.9 Environmental Protection Act 1990

The Environmental Protection Act 1990 and the Waste Management Licensing Regulations 1994 brought new and more stringent controls to all landfill sites and also applied a duty of care on anyone dealing with waste to do so in safe manner. The Act also allowed for the formation of the Environment Agency which was charged with regulating waste management (amongst other things). It introduced a more effective licensing system to replace the provisions of the

Control of Pollution Act 1974. It applies to all controlled wastes (household, commercial, industrial and special waste).

This lead to increasing regulation and monitoring of landfill site operations. All sites that were likely to generate leachate or gas had to be located in impermeable geology and/or lined with comprehensive barriers, capping layers and leachate/gas collection systems. The waste was covered daily with soils to reduce odour and other emissions during operation. Emissions (gas, vapours and leachate) were heavily monitored and safely managed and dilute and disperse was no longer an acceptable practice. Wider issues such as the effect of methane emissions on global climate change were now considered, for example many domestic refuse sites were carefully managed to optimise gas generation to allow its collection to power electricity generators.

It became much more difficult to surrender licences. This could only be done when it had been demonstrated the landfill leachate and gas no longer posed a serious risk to the environment. Thus the cost of landfilling needs to consider the post-closure maintenance and monitoring for tens or possibly hundreds of years. The cost of waste disposal has increased and will continue to increase.

The Landfill Tax was introduced in the UK in 1996 to try to reduce the amount of waste being sent to landfill. All waste placed in a landfill is subject to the tax unless specifically exempt. For example soil being removed from redevelopment sites because it is contaminated is exempt. The Special Waste Regulations 1996 updated the definition of special waste (hazardous waste) to conform to European Union legislation.

3.2.10 21st century

In 2000 the UK Government introduced the National Waste Strategy that set new targets for local authorities to achieve in recycling waste. Incineration and composting are being increasingly used to dispose of waste and thus the volume of waste being landfilled is starting to reduce.

In the late 1990s the Environment Agency began to apply the Waste Management Licensing Regulations much more rigidly to redevelopment sites where contaminated soil was being treated or moved around on site. This has resulted in some redevelopment sites being officially classified as landfill sites, even though the works have been undertaken to make the site suitable for development. There are some sites where contaminated materials have been buried in designated parts of the site such as landscaping areas that are classified as landfill sites.

There are even sites where excess clean spoil from motorway construction has been used to raise ground levels in farmland that have been subject to Waste Management Licensing and will be classified as landfill sites. Thus the nature of landfill sites can be very different depending on the type of waste they have

accepted and the nature of the gas and leachate management systems that have
been installed.

3.3 Ground gas and vapour properties

The properties of some of the most common ground gases and vapours that
need to be assessed to allow developments to proceed are given in Tables 3.2
and 3.3. This is not an exhaustive list but provides the most commonly required
properties when undertaking risk assessment of gases and vapours. Additional
information on trace gases is given in the guidance provided by the Environment Agency (2004a).

In this handbook ground gas and vapour are defined as follows:

- Ground gas: this is generated by processes such as biodegradation,
  chemical reactions etc.
- Soil vapour: this occurs as a result of volatilisation from contamination
  that has been spilt or been placed in the ground

Table 3.2 Useful properties of some hazardous ground gases (see also CIRIA Reports 130 and C659/665)

	Methane	Carbon dioxide	Hydrogen sulphide	Carbon monoxide	Ammonia	Hydrogen	Hydrogen cyanide
			Physical and chemical properties				
Hazard	Flammable, explosive	Toxic	Flammable, explosive, toxic	Toxic, flammable, explosive	Toxic, corrosive	Flammable, explosive	Toxic, flammable
Description	Colourless, odourless and flammable gas. Important greenhouse gas	Colourless, odourless and toxic gas. Important greenhouse gas	Colourless, flammable and toxic gas. Rotten eggs odour at low concentrations <1 ppm, but odourless at concentrations > about 50 ppm due to anaesthesia of olfactory sense	Colourless, toxic, odourless and flammable gas	Colourless, toxic gas with characteristic, irritating pungent odour. Burns in oxygen	Colourless, odourless and flammable gas	Colourless, toxic gas, faint bitter almond like odour. Explosive in air
Formation	Anaerobic degradation of organic material	Aerobic and anaerobic degradation of organic material, action of acid Water in carbonate rocks and respiration of soil bacteria	Sulphate-reducing bacteria obtain energy by oxidising organic matter or hydrogen with sulphates and Produce H_2S. Happens in low-oxygen environments, e.g. in swamps or standing waters. Other anaerobic bacteria produce H_2S by digesting amino acids containing sulphate	Incomplete combustion of organic materials. In landfill waste can also be produced By the reduction of carbon dioxide by nascent hydrogen	Fixation of atmospheric nitrogen by soil enzymes. Also from degradation of amino acids in Waste by soil bacteria	Chemical reaction between water and finely divided metal (e.g. aluminium) in the ground	Degradation of cyanohydrins in vegetables and fruits. Degradation of plastics Containing nitrogen. Can be synthesised by the reaction of methane and ammonia. Action of acid on organic cyanide salts

Table 3.2 Useful properties of some hazardous ground gases (see also CIRIA Reports 130 and C659/665) (continued)

	Methane	Carbon dioxide	Hydrogen sulphide	Carbon monoxide	Ammonia	Hydrogen	Hydrogen cyanide
Chemical symbol	CH_4	CO_2	H_2S	CO	NH_3	H_2	HCN
Molecular weight (atomic mass units)	16	44	34	28	17	2	27
Density (kg/m³)	0.71	1.98	1.53	1.25	0.68	0.085	0.687
Solubility in water at STP (mg/l)	25	1450	4100	21.4	899,000	1.62 (@21°C)	Completely miscible
Viscosity (Ns/m²)	1.03×10^{-5}	1.4×10^{-5}	1.0×10^{-5}	1.66×10^{-5}	9.8×10^{-6}	8.7×10^{-6}	Not reported
Diffusion coefficient in air (m²/s) (@STP)	1.5×10^{-5} m²/s	1.39×10^{-5}	1.76×10^{-5}	1.96×10^{-5} (@9°C)	1.98×10^{-5}	6.1×10^{-5}	Not reported
Hazardous Properties							
Lower explosive or flammable limit (% v/v in air)	5	Non-combustible	4.5	12.5	Non-combustible	4	5.6

Table 3.2 Useful properties of some hazardous ground gases (see also CIRIA Reports 130 and C659/665) (continued)

	Methane	Carbon dioxide	Hydrogen sulphide	Carbon monoxide	Ammonia	Hydrogen	Hydrogen cyanide
Upper explosive or flammable limit (% v/v in air)	15	Non-combustible	45.5	74.2	Non-combustible	74	40
Toxicity	Not toxic (but can cause asphyxiation by displacing oxygen)	Headaches and shortness of breath at 3% becoming severe at 5%. Loss of consciousness at 10%, fatal at 22%	At 20–150 ppm watering eyes, blurred vision, shortness of breath, sore throat. At 400–500 ppm pulmonary oedema, headache, dizziness, coma, asphyxiation	Symptoms of mild poisoning include: headaches and flu-like effects. Greater exposure can lead to loss of consciousness and death	Irritant to skin, eyes, throat, coughing, burns, lung damage, death	Non-toxic (but can cause asphyxiation by displacing oxygen)	Highly toxic by inhalation and skin contact resulting in nausea and death
Work place exposure limit	None provided	8 h at 5000 ppm 15 min at 15,000 ppm	8 h at 5 ppm 15 min at 10 ppm	8 h at 30 ppm 15 min at 200 ppm	8 h at 25 ppm 15 mins at 35 ppm (anhydrous ammonia)	Non-toxic. None provided	15 min at 10 ppm

Table 3.2 Useful properties of some hazardous ground gases (see also CIRIA Reports 130 and C659/665) (continued)

	Methane	Carbon dioxide	Hydrogen sulphide	Carbon monoxide	Ammonia	Hydrogen	Hydrogen cyanide
Environmental assessment levels for air (µg/m³) (Environment Agency, 2003b)	None provided	None provided	140 short-term 150 long-term	350 short-term 10,000 long-term	180 short-term 2500 long-term	None provided	3
Notes	Explosive limit changes when oxygen concentration reduces. When carbon dioxide concentration reaches 25% methane is non-flammable Oxidises to carbon dioxide by bacterial action		After short period of exposure paralyses sense of smell				Like hydrogen sulphide short exposure can result in the loss of sense of smell

[1] 1 ppm = 0.0001 % v/v

Table 3.3 *Properties of some commonly occurring vapours (see also CIRIA Report C659/665)*

	Benzene	Toluene	Xylene (mixed isomers)	Naphthalene	Tetrachloroethene (PCE)	Trichloroethene (TCE)	Vinyl chloride (VC)
				Physical and chemical properties			
Hazard	Flammable, toxic	Flammable, toxic	Flammable, toxic	Flammable, toxic	Toxic	Toxic	Flammable, toxic
Description (at STP)	Volatile, colourless liquid, sweet odour	Volatile, colourless liquid, sweet odour	Volatile, colourless liquid, sweet odour	Crystalline, aromatic, white, solid hydrocarbon. Odour of mothballs	Volatile, colourless liquid with mildly sweet odour. Also known as tetrachloroethylene, PERC, perchloroethylene, and PCE.	Colourless volatile liquid with sweet odour like chloroform. Vapour is much denser than air. Also known as trichloroethylene, ethylene trichloride, TCE and trilene	Colourless gas with slightly sweet odour
Formation	Fuel spills are main source in ground. Also present in cigarette smoke	Fuel spills are main source in ground	Fuel spills are main source in ground,	Fuel spills, gas works	Chemical spills: used widely in dry cleaning and degreasing	Chemical spills: used widely in cleaning and degreasing, degradation of PCE	Degradation of PVC/TCE and related polymers and certain chlorinated Solvents
Chemical symbol	C_6H_6	C_7H_8	C_8H_{10}	$C_{10}H_8$	C_2CL_4	C_2HCL_3	C_2H_3CL
Molecular weight	78.1	92.4	106.2	128.2	165.8	131.4	62.5
Solubility in water at STP (mg/l)	1750	515	198	31	200	1000	2700
Viscosity (Ns/m²)	6.52×10^{-4}			8.8×10^{-4}			

Table 3.3 *Properties of some commonly occurring vapours (see also CIRIA Report C659/665) (continued)*

	Benzene	Toluene	Xylene (mixed isomers)	Naphthalene	Tetrachloroethene (PCE)	Trichloroethene (TCE)	Vinyl chloride (VC)
Diffusion coefficient in air (cm²/s)	0.088	0.085	0.072	0.059	0.072	0.082	0.106
Hazardous properties							
Lower explosive or flammable limit (% v/v in air)	1.2	1.1	1.1 (paraxylene)	0.9	Not combustible	Not combustible	3.8
Upper explosive or flammable limit (% v/v in air)	7.1	7.1	6.6 (paraxylene)	5.9	Not combustible	Not combustible	31
Toxicity	Causes drowsiness, dizziness and unconsciousness, carcinogen	Causes drowsiness, dizziness and unconsciousness, carcinogen	Causes dizziness, confusion and loss of balance	Causes headaches, confusion, excitement, nausea and vomiting. May be dyseria, haematuria and acute haemolytic reaction. Carcinogen	Vapour causes irritation of the eyes, nose and throat. Suppresses central nervous system. High concentrations cause unconsciousness and death. Carcinogen	Vapour causes irritation of the eyes, nose and throat. Suppresses central nervous system. High concentrations cause unconsciousness and death. Carcinogen	Vapour causes dizziness and sleepiness. High concentrations lead to unconsciousness and death. Carcinogen
Workplace exposure limits	8 h at 1 ppm	8 h at 50 ppm 15 min at 150 ppm	8 h at 50 ppm 15 min at 100 ppm	None provided	8 h at 50 ppm 15 min at 100 ppm	8 h 100 ppm 15 min at 150 ppm	8 h at 3 ppm

Table 3.3 *Properties of some commonly occurring vapours (see also CIRIA Report C659/665) (continued)*

	Benzene	Toluene	Xylene (mixed isomers)	Naphthalene	Tetrachloroethene (PCE)	Trichloroethene (TCE)	Vinyl chloride (VC)
Environmental assessment levels for air ($\mu g/m^3$)[1]	Long-term 16.25 Short-term 208	Long-term 1910 Short-term 8000	Long-term 4410 Short-term 66200	Long-term 530 Short-term 8000	Long-term 3450 Short-term 8000	Long-term 1100 Short-term 1000	Long-term 159 Short-term 1851
Sorption coefficient Log K_{oc} (log l/kg)	1.77	2.13	2.38	3.30	2.19	2.10	1.75
Henry's law constant	0.2289	0.2600	0.2900	0.0199	0.7588	0.4136	0.0560
Approximate odour threshold (ppm)	1.5	8	1	0.3	47 Note most people stop noticing odour after a short period	50	4000

[1] From Environment Agency (2003b). These values are updated from time to time. Check the Environment Agency website for the most up to date values

SUMMARY: Sources and properties of ground gas and vapours

There are numerous man-made or natural sources of ground gas including landfill sites. Gas can also be trapped in material such as coal and be released when excavations for mines take place (or in peat and released when activities such as piling occur). The level of risk associated with differing sources varies according to how much gas can potentially be generated. The various sources and typical level of risk associated are given in this chapter.

Landfilling has been used as a means of waste disposal throughout history and over time the nature of wastes and the engineering of landfills has changed. Thus the age of the made ground or landfill will determine to some extent the likely risk it poses with respect to ground gases. Any general assessment of the risks posed by the various sources can only provide an initial indication of the likely risks. An initial assessment made using the data in this chapter should always be confirmed by site investigation.

Landfill sites that have been filled predominantly with industrial waste are difficult to assess without a detailed knowledge of what materials have been placed in them.

This chapter also provides data on the physical and hazardous properties of gases and vapours that are required to undertake the more detailed risk assessment or mathematical modelling of gas and vapour flows that will be described later in this handook.

The list of data is by no means exhaustive and further information can be obtained from a number of sources:

- Health and Safety Executive
- Environment Agency

CHAPTER FOUR

Gas and vapour generation and migration

This Chapter discusses the generation of various ground gases or the production of vapours. It focuses mainly on the generation of the bulk gases of methane and carbon dioxide by the biodegradation of organic material. This can occur in any soils where organic material is present, and especially in landfill waste, to produce landfill gas. It also occurs where there are hydrocarbons in the ground (e.g. diesel and petrol spills or leaks), although in these cases a low volume of gas will be generated.

The majority of natural methane and carbon dioxide in peat and alluvium has already been generated and is largely trapped in the soil. It therefore represents a finite source and the likely risk is related to the volume present in the ground and factors that could affect surface emissions such as variations in groundwater levels causing a pumping effect.

Methane in coal is only released in significant volumes when a void is made in the coal (e.g. workings or a borehole). The methane then desorbs from the coal into the void. This can continue for up to 50 years. Chemical and biological oxidation of methane then forms carbon dioxide. Coal bed methane can also desorb into open fissures and faults.

4.1 Methane and carbon dioxide generation

To understand how methane and carbon dioxide are produced a knowledge of the biochemical and microbiological breakdown of organic matter in soil or landfill waste is required. A simple summary of the process is shown in Figure 4.1.

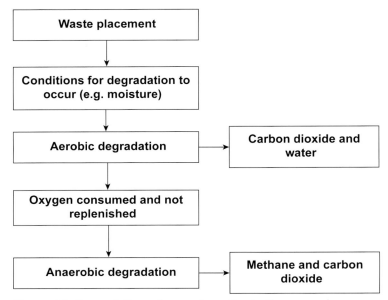

Figure 4.1 *Degradation of organic matter in the ground*

Immediately after tipping the organic matter is in an aerobic state, i.e. air pockets exist in the soil or waste, so the initial decomposition of the organic matter is an aerobic biological process and the two main end products are carbon dioxide and water. This process also occurs during the composting of waste.

Bacteria do not flourish in dry conditions, so biodegradation starts only when the soil or waste is moist. Once aerobic biodegradation has commenced oxygen is soon depleted and if no replenishment is available, the material becomes anaerobic. Anaerobic processes use oxygen present in the structure of the organic matter or from inorganic ions such as sulphate. The main products from anaerobic degradation are methane and carbon dioxide. A simple summary of the process is provided in Box 4.1.

Box 4.1 Chemistry of biodegradation

Aerobic processes

In very simple terms aerobic processes can be summarised as:

$$\text{organic substances} + \text{nutrients} + O_2 \rightarrow CO_2 + H_2O + NO_3^- + PO_4^{3-} + SO_4^{2-} + \text{new cells} + \text{energy}$$

(C, O, H, N, P, S)   (N, P etc.)

An example of the aerobic biochemical oxidation of acetic acid is thus:

$$CH_3COOH + 2O_2 + \text{bacteria} \rightarrow 2CO_2 + H_2O$$

Anaerobic process

A generalised anaerobic process can be summarised as:

$$\text{organic substance} + \text{nutrients} \rightarrow CH_4 + CO_2 + N_2 + PH_3 + H_2S + \text{new cells} + \text{energy}$$

(C, O, H, N, P, S) (N, P etc.)

An example of the anaerobic biochemical oxidation of acetic acid is:

$$CH_3COOH + \text{bacteria} \rightarrow CH_4 + CO_2$$

Anaerobic decomposition takes place in two phases as follows:

- Complex organic compounds, such as cellulose and lignin, are broken down into simpler salts and acids, typified by acetic acid, proprionic acid and pyruvic acid or alcohols. This process is known as acidification
- Methanogenic bacteria utilise the end products from the first phase and yield methane and carbon dioxide. This is also known as methanogenesis

Methanogenesis is a complex process with biochemical degradation occurring at widely varying rates of reaction and states of chemical decomposition depending on the nature of the waste or organic soil. There are, however, some important factors that affect the rate of gas generation and the total volume of methane and carbon dioxide that is produced. These factors should be considered when assessing the risks posed by a source (see Box 4.2).

Box 4.2 Factors affecting generation of methane and carbon dioxide

Volume of degradable material
To generate large volumes of methane and carbon dioxide a large mass of readily degradable organic content is required (e.g. domestic landfill sites). The volume of gas generated by other sources depends on the volume of degradable material that is present in the soil and the total volume of the source (e.g. made ground, dock silt, compost heaps, buried topsoil etc.).

Foundry sands frequently contain organic materials such as phenolic binders, coal dust and pieces of wood and paper. However, the overall volume of degradable material is usually low and the volumes of gases generated are correspondingly low.

Moisture content
Moisture content affects gas generation. Idealised generation profiles assume that the waste is at an optimal moisture content. Generally, the higher the

moisture content the higher the rate of degradation (although rates reduce when saturation is approached). If waste is dry it could have been in the ground for many years and not reached peak generation rates. In that case the risk of future moisture content changes would need to be assessed (i.e. will it become wet and rates increase). Gas generation can still occur below the groundwater or leachate head but at lower rates. Significant generation can occur below a perched water table in material below the aquiclude.

Nature of degradable material

Woody materials usually have slow rates of gas generation (rates of generation are low but over a lengthy period of time) because the presence of lignin reduces anaerobic degradability (Pueboobpaphan and Toshihiko, 2007). Some organic materials are effectively non-degradable (e.g. coal and newspaper are high in lignin). The most readily degradable materials contain cellulose or simple sugars/starches (e.g. food waste and commercially produced paper). If the degradable material is finely shredded (so it has a large surface area) it will degrade more quickly (e.g. a log will degrade very slowly but if it is turned into sawdust it will degrade at much faster rate) (see also Section 5.9).

pH and other conditions

The pH value of the waste or soil affects the rate of gas generation. Optimum conditions for methanogenesis are pH 6.5–8.5. Highly alkaline conditions will inhibit gas generation (e.g. if high quantities of lime are present). The rate of degradation can also be severely inhibited by the presence of many materials that are toxic to the micro-organisms involved (e.g. heavy metals).

Lining and capping

The nature of the lining and capping of an engineered landfill will affect the rate of landfill gas generation. The capping layer affects the rate of rainfall infiltration and thus moisture conditions. Lining layers will affect the build-up of leachate within a landfill, which again will affect the rate of gas generation.

Age of made ground or landfill

If conditions are suitable, the maximum methane and carbon dioxide generation will occur in the first 10–15 years after filling. After this period the gas generation will decline to a much lower residual level (see Section 4.2). It is important to realise that in some cases gas generation may not have started because it is dry, even though the waste has been in the ground for a long while. One site was underlain by significant deposits of sawdust within made ground that had not degraded and thus represented a high potential source, even though the waste had been in the ground for over 20 years.

4.2 Estimating the volume of methane and carbon dioxide that is generated

An initial approximation of the landfill gas generation can be produced by simply assuming that each tonne of fresh biodegradable waste will produce 10 m³ of methane per year (UK landfills typically generate 5–10 m³/t/yr). The following calculation will then give an approximation of the rate of methane generation (Environment Agency, 2004a). This equation will produce an overestimate of gas flow at peak production from historic waste or organic soils.

$Q = M \times 10 \times T/8760$

Where:

Q = methane flow (m³/h) from fresh waste

M = annual quantity of biodegradable waste (tonnes)

T = time in years during which waste has been placed.

Thus MT = total quantity of biodegradable waste placed over lifetime of landfill.

This is a very simplistic analysis and makes no allowance for reducing rates of generation with time, but it is useful as a first estimation of gas generation to assess the potential scale of a problem. A graph of landfill gas generation (assumed to be the combined total of methane and carbon dioxide) for fresh waste was proposed by Nastev (1998). This also estimates the declining rate of generation with time (Figure 4.2).

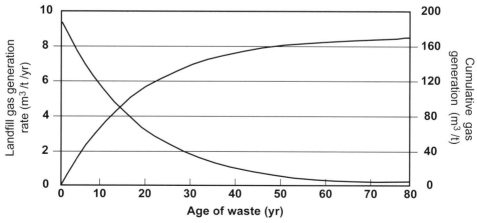

Figure 4.2 *Landfill gas generation against time (Nastev, 1998)*

A more accurate estimation of landfill gas generation rates and the changes over time can be developed using the GasSim computer program developed by the Environment Agency (see Figure 4.3). This uses a first-order kinetic model to estimate gas generation (i.e. exponential decline), with no lag or rise period, and with waste fractions categorised as being of rapid, medium or slow degradability. This equation (or similar first-order equations) is commonly used in combination with waste input predictions to produce a gas generation profile for the lifetime of the site. Such a multi-phase, first-order decay equation forms the core of the GasSim model (Environment Agency, 2002c).

The equation used in GasSim is:

$$\alpha_t = 1.0846.A.C_i.k_i.e^{-k_i t}$$

Where:

α_t = gas generation rate at time t (m³/yr)

A = mass of waste in place (tonnes)

C_i = carbon content of waste (kg/tonne)

K_i = rate constant (yr⁻¹) (0.185, fast; 0.100, medium; 0.030, slow)

t = time since deposit (yr)

The graph in Figure 4.3 shows how the total landfill gas generation rate computed from GasSim varies over time. In this example tipping was undertaken in the period 1977–1992 at which date peak gas generation is indicated to occur. By 2006 the generation rate has fallen to about 40% of the peak rate.

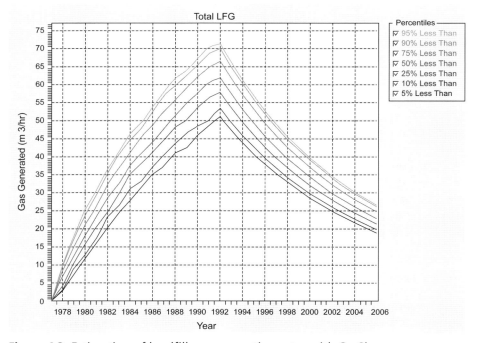

Figure 4.3 *Estimation of landfill gas generation rates with GasSim*

However, on development sites it is usual for significant assumptions to be made regarding the length of time over which filling occured and the progress of capping over time. Similarly,the nature of the material placed below a site and how degradable it is can only be estimated on the basis of on-site investigation. Thus, although predictive generation curves are a useful tool, the associated uncertainty should be clearly understood.

The nature of a landfill site or other source such as made ground will affect the nature of the decomposition (aerobic or anaerobic) and thus the volume of methane and carbon dioxide that is generated. Unmanaged landfills produce less methane from a given amount of waste than managed landfills, because a larger fraction of waste decomposes aerobically in the top layers of unmanaged landfills. In shallow sources such as made ground that is less than 5 m deep this effect is even greater (United Nations Convention on Climate Change (UNFCCC) 2005). Specific testing can be undertaken to determine the volume of material in a soil or waste that will degrade to produce gas (see Section 5.9).

4.3 Generation of minor component gases (trace gases)

Environment Agency guidance on the management of landfill gas (Environment Agency, 2004a) states that other gases can also be generated from non-biodegradable (commonly referred to as inert) wastes by the following chemical processes:

- Corrosion of metals or reactions between metals and alkali wastes, e.g. hydrogen is emitted when zinc or aluminium is attacked by alkalis
- Formation of gases by reaction of waste with acids, e.g. hydrogen sulphide released by the reaction of acids with sulphides, commonly found in gasworks waste, or hydrogen cyanide released by the reaction of acids on cyanides found in metal plating or refining residues
- Release of free bases by reaction of the waste with alkali, e.g. amines are emitted
- Redox reactions within the waste, e.g. sulphur oxides are emitted

Minor components of landfill gas will be affected by similar factors such as methane and carbon dioxide generation as described in Section 4.1. The potential to generate component gases depends on the specific 'cocktail' of the various sources deposited in the waste, the interaction of gases themselves and the chemical and biochemical interaction between soil, water and the waste.

Other gases that can pose a risk and are, therefore, of concern include:

- Hydrogen sulphide
- Hydrogen cyanide
- Carbon monoxide
- Ammonia
- Hydrogen

The generation of these gases is discussed below.

Hydrogen sulphide: it is possible to obtain high concentrations of hydrogen sulphide if large amounts of sulphate bearing material (e.g. plaster board containing gypsum (calcium sulphate)) are disposed jointly with degradable material. Under anaerobic conditions sulphates are reduced to sulphides by bacterial action. Hydrogen sulphide is released by the action of acidic ground-water water or leachate on the sulphide compound. It can also be present in sewage sludge and where metal waste slag is present. Sulphides also occur naturally in peat, typically as iron sulphide.

Hydrogen cyanide: cyanide salts are widely used in industry for example metal plating and engraving, dyes and printing inks as well as many important plastics and chemicals. They are also common in slag wastes and ash. Hydrogen cyanide can be released by the action of acids on cyanide salts. The reaction is typically:

$$H^+ + NaCN \rightarrow HCN + Na^+$$

It is also possible for hydrogen cyanide to be synthesised from methane and ammonia by the reaction:

$$CH_4 + NH_3 \rightarrow HCN + 3H_2$$

This reaction normally requires a high temperature and a catalyst for synthesis to take place but it can occur as a result of the action of enzymes or soil bacteria acting as a biocatalyst.

Carbon monoxide: is produced by the partial combustion of organic material in the absence of free oxygen. Large volumes can be generated by underground fires in coal mines and in landfill sites where incomplete combustion results in carbon monoxide being formed in preference to carbon dioxide. It can also be produced in low volumes by the reduction of carbon dioxide if metals such as aluminium or zinc are attacked by alkalis to produce nascent hydrogen that in turn reduces carbon dioxide to form carbon monoxide and water:

$$2H^+ + CO_2 \rightarrow CO + H_2O$$

Ammonia: is produced by the reduction of nitrogen-containing matter (e.g. aminoacids, amines, amides and nitrates) in soil to form ammonia. For example in organic rich soils urea can be hydrolysed to form ammonia and carbon dioxide:

$$CON_2H_4 + H_2O \rightarrow CO_2 + 2NH_3$$

Decomposition of nitrogen compounds involves chemical and biochemical processes (protein and bacteria in a landfill). In the presence of hydrogen sulphide, a powerful reducing gas, nitrates can be reduced to nitrites and these latter compounds are themselves reduced in anaerobic environments and converted to ammonia. The generalised equation is:

$$HNO_3 + H_2S \rightarrow NH_3 + SO_3$$

In this example the SO_3 readily combines with water in the landfill to form sulphates:

$$SO_3 + H_2O \rightarrow H_2SO_4$$

Similar reactions can occur in landfill waste by the reduction of nitrates by nascent hydrogen:

$$9H^+ + NO_3^- \rightarrow NH_3 + 3H_2O$$

Hydrogen: if metals are placed in water and an electrode potential is formed then a cathodic reaction can produce nascent hydrogen which is a very powerful reducing agent. Typically, zinc in water will have a tendency to form positive charged zinc ions in aqueous solution. If an electrochemical cell is formed hydrogen can be liberated from the aqueous solution as follows:

$$Zn + 2H^+ \rightarrow Zn^{2+} + H_2$$

Normally this will be in small volumes, although there has been an example where such a reaction occurred in a steel tube inserted in the ground. Hydrogen accumulated inside the steel tube (a small volume) and when it was cut the sparks triggered an explosion. Generation of hydrogen is also associated with finely divided aluminium waste deposited in landfill.

4.4 Vapour production

Vapours can be produced in landfills but probably the most common cause of vapours in the ground is the presence of hydrocarbon contamination from fuel and chemical spills or leaks in locations such as petrol filling stations or industrial sites. In these cases the vapours are present as a result of volatilisation from contamination of the soil or groundwater.

In landfill gas generated from landfill sites there are several hundred trace volatile hydrocarbon gases that can be present, for example volatile organic compounds (VOCs). Their total presence is typically about 1% of the total volume of landfill gas. There are many household chemical products that will contain these compounds. Thus, in certain circumstances, elevated concentrations of VOCs might be detected.

Volatile organic compounds may be released from waste by the following physical processes:

- Gas stripping with other released gas or water vapour
- Heat generated in the waste
- Aerosols carrying liquid droplets
- Dust carrying sorbed materials
- Reactions between organic compounds to form more volatile species, e.g. the formation of esters

In general, vapour generation will be at much lower rates than the bulk gases of methane and carbon dioxide, but they pose health risks at much lower concentrations. More detailed guidance on monitoring trace gases has been given by the Environment Agency (Environment Agency, 2004a).

The source of vapours (hydrocarbon contamination) can also migrate significant distances in groundwater. However, this type of migration is beyond the scope of this handbook and readers should refer to the wealth of other books that cover this subject.

4.5 Gas and vapour migration

Gas and vapours migrate by pressure driven flow and/or diffusive flow through the soil pore spaces. Migration can also be driven by the buoyancy of gases. Migration of a vapour source in groundwater (i.e. the dissolved contamination) can be a significant pathway for vapours to reach off-site receptors on many sites.

The bulk gases (methane and carbon dioxide) can also migrate in groundwater. Methane is slightly soluble in water (25 ml methane/litre water at STP). Carbon dioxide is more soluble in water, ionising to form bicarbonate and carbonate ions. This difference in solubility is one of the factors responsible for observed variations in bulk gas composition in monitoring wells.

Taking methane as an example the volume dissolved in water depends on the partial pressure. The mass of methane dissolved in groundwater at 10°C is given by:

$M = 29.9 \times P$ (from Hooker and Bannon, 1993)

Where:

M = mass of dissolved methane (mg/l)

P = partial pressure of methane (atm)

An example calculation of the volume of gas that can migrate in this way is given in Box 4. 3.

Box 4.3 Estimate of methane migration in groundwater (Hooker and Bannon, 1993)

For methane to come out of solution from the water there needs to be a change in pressure.

Assume the water is pumped from an aquifer where the partial pressure is equal to the hydrostatic pressure of 3 atm (304 kPa or 30 m head of water):

$$M = 29.9 \times 3 = 89.7 \text{ mg/l}$$

When it comes to the surface the partial pressure will come into equilibrium with the air and the partial pressure will be 1.6×10^{-6} atm so the volume of gas in the water is:

$$M = 29.9 \times 1.6 \times 10^{-6} = 4.78 \times 10^{-5} \text{ mg/l}$$

i.e. 89.7 mg/l is released (or virtually all the methane). This demonstrates that migration in groundwater is not likely to occur unless there are large changes in the partial pressure of the methane. This is unlikely in shallow landfill and similar sites (up to 5 m depth) where there is no great variation in groundwater levels. Migration via this route will become more likely from a deep landfill with a high leachate head and a confined migration pathway for the groundwater, or from deep geological sources.

4.6 Modelling gas and vapour migration

The extent and rate of gas migration out of the ground or along in ground pathways depends on three key factors:

- Generation rate of gas and build up of pressure or a concentration gradient
- Presence of a permeable (i.e. preferential) gas migration pathway
- Buoyancy of the mixture

If migration pathways are present, gas will migrate either vertically or horizontally through the ground to an exit point. Depending on the specific site conditions, migration can be driven by diffusion or pressure. Small quantities of trace gases or vapours within the landfill can be transported effectively by the migration of landfill gas acting as a carrier gas.

In order for sustained migration to occur, the gas must be replenished at source. As such significant gas generation is usually required for large volumes of gas to migrate by either diffusive or advective flow. The volume of the gas is much more important than the concentration of the gas in the ground.

The effect of gas volumes and pressures is illustrated in Figure 4.4, which shows an example of two balloons filled with methane. The balloon on the left-hans side can be compared to a large domestic landfill producing large volumes of gas. If a leak is present the gas will rush out of the balloon. The one on the right-hand side is comparable to many brownfield sites under-lain by predominantly inert made ground generating limited volumes of gas. There is gas in the balloon but if there is a leak it will come out very slowly, and the supply is much more limited.

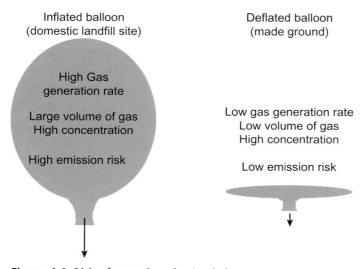

Figure 4.4 *Risk of gas migration/emission*

Gas concentration within a source is a poor indicator of migration or surface emission risk. The methane concentration could be measured at similar levels in both balloons but the risk of migration is obviously higher in the left-hand one. Even where migration can occur via diffusion there needs to be a con-stant source of gas to replenish the gas that migrates, in order to maintain a concentration gradient.

The first step in modelling gas migration (either horizontal or vertical) is to develop a conceptual model for the site that identifies the potential migration pathways. Examples of migration pathways are shown in Figure 4.5. Migration pathways can be along natural routes such as permeable layers of sand and gravel or along faults and fissures. Man-made migration routes can be important. For example gas can migrate along buried services or the instal-lation of stone columns or piles to provide foundations can form a pathway through an upper impermeable layer that seals the ground gas in a lower stratum.

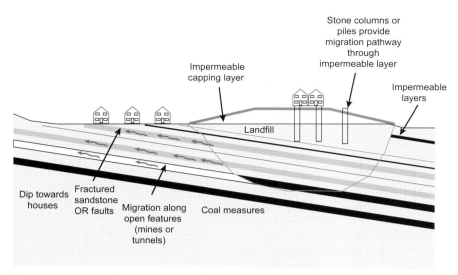

Figure 4.5 *Examples of gas migration routes*

The geological conditions in and around a site have a major influence on the risk of gas migration. A commonly used value of 250 m is often incorrectly perceived as being the limit to which gas migration can occur. Migration has been observed to occur for distances up to 400 m along open features (e.g. open faults or other similar features). Conversely there have been sites where no migration has been observed as close as 10 m to sources that are surrounded by impermeable clay (e.g. London clay).

Methane from mineworkings usually migrates to the ground via fissures in the overlying rock or via shafts and adits. Emissions of methane (fire damp) or carbon dioxide (black damp) from the ground are a common problem in areas where coal measures outcrop near the ground surface or adits and shafts connect with abandoned deep underground workings. The cessation of mine drainage can result in the rapid expulsion of methane and carbon dioxide from abandoned workings as groundwater levels rise creating a 'piston effect'.

A simple mathematical assessment of gas migration can be achieved using two basic models:

- Pressure driven flow (also known as advective flow) that is modelled using Darcy's law
- Diffusive flow that is modelled using Fick's law

More detailed modelling of gas migration can be undertaken using finite element methods. However, the degree and quality of data that is required for effective use of such numerical methods is rarely obtained for typical development projects and many of the parameters have to be estimated, which limits the usefulness of the analysis. Conversely a very crude estimate of the vertical migration of gas from the ground can be obtained using measured borehole flow rates with an empirical relationship proposed by Pecksen (1986).

The mathematical models used to estimate the rate of gas flow through the ground are very simplified and assume that there is no change in gas composition as it migrates through the ground. For example they do not take account of the fact that methane may oxidise as it passes through oxygen-rich near-surface soils. It is also important to realise that some of the parameters used in the equations can show a high degree of variation and therefore a sensitivity analysis is often required when modelling gas migration. For example soil, and in particular made ground, rarely has homogeneous properties and as such permeability can vary considerably.

The models also assume steady-state conditions and that gas flow behaves as a liquid (i.e. the gas or vapour is incompressible). These simplifying assumptions introduce uncertainty into the results of any modelling. However, as long as such factors are taken into account, simplified mathematical models are a useful aid to decision making and can act as a check on the results of more complex mathematical models.

4.6.1 Pressure driven flow

This can be modelled simply using Darcy's law of fluid flow through porous media. The equation can be used to model both horizontal and vertical gas flow. It can therefore be used to estimate gas surface emission rates.

4.6.2 Darcy's law

The equation for Darcy's law is:

Flow of gas being considered in ground, $Q_v = \left[\dfrac{K_i \gamma A i}{\mu}\right] \times$ gas or vapour concentration

Where:

Q_v = flow of gas being considered, in m^3/s through area A

K_i = intrinsic permeability of material through which gas or vapour is flowing (m^2)

γ = unit weight of gas (N/m^3)

μ = viscosity of gas being considered (Ns/m^2)

A = area of migration perpendicular to migration direction (m^2)

i = pressure gradient along migration route (as fluid gradient for gas considered) = (gas pressure/unit weight)/length (Pa)

Typical gas properties are given in Tables 3.2 and 3.3.

An example of how the equation is applied is given in Box 4.4. Darcy's law is also used in the equations quoted by Johnson and Ettinger (1991) to estimate vapour migration from the ground into a building. An example showing the application of Darcy's law to estimate vertical emissions of gas from the ground is provided later in this chapter (see Box 4.7).

The Darcy equation is based on the intrinsic permeability of the soil or rock through which the gas is flowing. It differs from the more common equation used to express Darcy's law which is specific to the flow of water through the ground, and is based on the saturated hydraulic permeability of the ground. It should be noted that the theory assumes laminar, uncompressible gas flow and does not take into account any chemical or biochemical reactions that may occur along the migration route.

The intrinsic permeability of a soil or rock is given by:

$$K_i = \frac{K_{Darcy}\mu}{pg}$$

Where:

K_i = intrinsic permeability of material through which gas or vapour is flowing in m^2

K_{Darcy} = hydraulic permeability of material through which gas or vapour is flowing in m/s

μ = dynamic viscosity of water = 1.002×10^{-3} Ns/m² (@ 20°C)

ρ = density of water = 1000 kg/m³

g = gravitational constant = 9.81 m/s²

Box 4.4 Example of gas migration calculation – pressure driven

Estimate the rate of methane gas migration over a distance of 100 m given the following parameters:

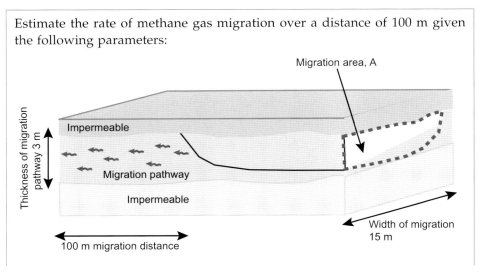

K_{Darcy} of soil migration pathway $= 1 \times 10^{-5}$ m/s

Therefore $K_i = 1 \times 10^{-5}$ m/s $\times 1.02 \times 10^{-7} = 1.02 \times 10^{-12}$ m^2

γ methane $= 7.2$ N/m^3

μ methane $= 1.03 \times 10^{-5}$ Ns/m^2

Pressure within source is 150 Pa (1 Pa $= 1$ N/m^2)

Thickness of soil migration pathway $= 3$ m

Width of gas migration front $= 15$ m

Therefore area of migration, $A = 15 \times 3 = 45$ m^2

i = pressure gradient along migration route (as a fluid gradient) = (gas pressure/unit weight)/length $= (70$ N/m^2/7.2 N/m$^3)/100$ m $= 0.097$

Q = flow of methane in m^3/s through area A $= \left[\dfrac{K_i \gamma A i}{\mu} \right] \times$ methane concentration

$$Q = \left[\frac{1.02 \times 10^{-12} \times 7.2 \times 45 \times 0.097}{1.03 \times 10^{-5}} \right] = 3.1 \times 10^{-6} \text{ m}^3/\text{s}$$

Q is the volume of methane migrating from the source through this pathway.

It should be noted that this assumes that there are no changes to the chemical composition of the gas during migration (e.g. oxidation of methane).

4.6.3 Diffusive flow

Diffusion depends on a concentration gradient being present (the gas flows from an area of high concentration to an area with a lower concentration). Molecular diffusion can be modelled using a number of laws but most of these are complex. If it is assumed that isobaric conditions are prevalent and there is no pressure driven flow, combined with relatively low concentrations of the gas being considered and high soil permeability, then Fick's law can be used.

4.6.4 Fick's law

Fick's law is also a component of the model used by Johnson and Ettinger to model gas or vapour migration into buildings. According to Fick's law the rate of mass transfer of a gas or vapour by diffusion can be estimated as follows (USEPA, 2003):

Diffusive rate of mass transfer of gas, $E = A(C_{source} - C_{g0})D^{eff}/L$

Where:

E = rate of mass transfer due to diffusion (g/d)

A = area through which migration occurs (m^2)

C_{source} = concentration of gas being considered at source (g/m³) (see Box 4.5)

C_{g0} = concentration of gas being considered at limit of migration (g/m³) (see Box 4.5)

D^{eff} = effective diffusion coefficient in the medium being considered (m²/d)

L = distance over which migration occurs (m)

Diffusion can be estimated through both the capillary zone and the unsaturated zone using different values for the effective diffusion coefficient.

This is the equation used in the Johnson and Ettinger (1991) model for vapour diffusion in the ground. Values of the diffusion coefficient of methane for various soils and rocks have been given in CIRIA Report 152 (O'Riordan and Milloy, 1995).

Alternatively, the effective diffusion coefficient is given by (USEPA, 2003):

$$D_{eff} = D^{air} \frac{\theta_v^{3.33}}{\theta_T^2} + \left(\frac{D^{H2O}}{H_i}\right) \frac{\theta_m^{3.33}}{\theta_T^2}$$

Where:

H_i = chemical-specific Henry's law constant (µg/m³−vapour)/(µg/m³ − H₂O)

θ_m = volumetric moisture content (m³ − H₂O/m³ − soil)

θ_T = total porosity (m³ −voids/m³−soil)

θ_V = volumetric vapour content (= $\theta_T − \theta_m$) (m³−vapour/m³−soil)

D^{air} = chemical-specific molecular diffusion coefficient in air (m²/d)

D^{H_2O} = chemical-specific molecular diffusion coefficient in water (m²/d)

Values for the parameters can be obtained from various references on the Johnson and Ettinger model (e.g. USEPA, 2003). An example showing the application of the equations is provided in Box 4.5.

It should be noted that Fick's law excludes the effects of Knudsen diffusion and non-equimolecular diffusion. It also assumes that diffusion only occurs in one direction and that the diffusion coefficient is constant (in practice it varies with concentration and temperature).

Figure 4.6 shows how the analyses are used in the Johnson and Ettinger method for vapour migration into buildings. There are numerous references available that explain how to use the Johnson and Ettinger method for vapour migration into buildings. However, one needs to be aware that the method assumes that vapour entry is via a basement (not that common in new UK developmentws) and that entry is only via a perimeter crack around the floor slab. This may not be the case in all buildings.

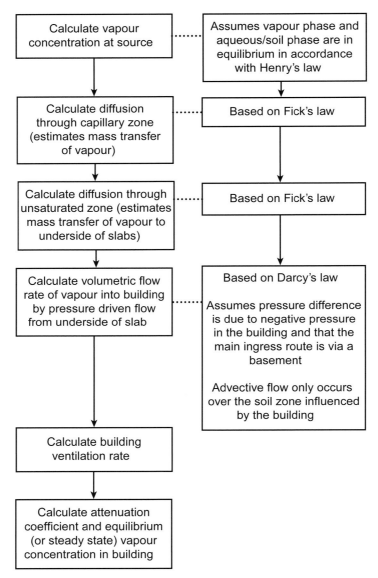

Figure 4.6 *Flow chart for Johnson and Ettinger method of vapour migration*

Box 4.5 Example of gas or vapour migration calculation – diffusion driven

Consider the diffusive flow of benzene over a distance of 100 m from a source concentration of 5000 ppm to a destination concentration of 10 ppm. The migration is through a layer of silty sandy gravel that is 3 m thick and the migration front is 15 m wide.

A gas or vapour concentration that is given in ppm can be converted to a volumetric measure using the following equation:

$C_v = C_{ppm}$ (12.187 × MW)/1000(273.15 + T) at an ambient pressure of 1 atm

Where:

C_{ppm} = concentration of gas or vapour (ppm)

C_v = concentration of gas (g/m³)

T = ambient temperature (°C)

MW = molecular weight of the gas or vapour (amu)

Therefore in this case for benzene 5000 ppm = 16.8 g/m³ and 10 ppm = 0.034 g/m³

θ_m = volumetric moisture content = 0.054 m³– H₂O/m³– soil

θ_T = total porosity = 0.375 m³–voids/m³ – soil

θ_V = volumetric vapour content (= $\theta_T - \theta_m$) = 0.375 – 0.054 = 0.321 m³ – vapour/m³ –soil

Assume that:

H_i = chemical-specific Henry's law constant = 0.2289 (µg/m³ – vapour)/ (µg/m³– H₂O) – this is actually unitless as it refers to the ratio between the aqueous phase concentration and the gas phase concentration.

D^{air} = chemical-specific molecular diffusion coefficient in air = 0.088 cm²/s = 76.03 m²/d

D^{H_2O} = chemical-specific molecular diffusion coefficient in water = 9.8 × 10⁻⁶ cm²/s = 0.0085 m²/d

So $D_{eff} = D^{air} \dfrac{\theta_v^{3.33}}{\theta_T^2} + \left(\dfrac{D^{H_2O}}{H_i}\right) \dfrac{\theta_m^{3.33}}{\theta_T^2}$

$D_{eff} = 76.03 \dfrac{0.321^{3.33}}{0.375^2} + \left[\left(\dfrac{0.0085}{0.2289}\right) \dfrac{0.054^{3.33}}{0.375^2}\right] = 12.2$ m²/d

Diffusive rate of mass transfer (g/d):

$E = A(C_{source} - C_{go})D^{eff}/L$

$E = 3 \times 15(16.8 - 0.034) 12.2/100 = 92.0$ g/d benzene migration

The use of these equations must be appropriate to the migration pathways identified in the conceptual model. Factors such as secondary permeability (e.g. via fissures in clay) and the effect of moisture content on the permeability to gas of soils should also be taken into account. The calculations also assume steady-state conditions, which are unlikely to occur in practice, and thus the results should only be used as an aid to judgement during any decision-making process.

4.7 Estimating surface emission rates

To model the surface emission rates of a gas or vapour an analysis of the vertical flow of gas in the ground is required. The most commonly used method to estimate surface emission rates is based on the simple assumption that a 50 mm borehole has a radius of influence of 1.78 m^2 which is equivalent to a surface area of 10 m^2 (Pecksen, 1991). This radius of influence was an arbitrary value chosen to ensure that surface emissions were not underes-timated. It is important to understand that the area of 10 m^2 applies to the surface area surrounding the borehole at ground level and is an estimate of the area of emission of gas from a single borehole. It should not be confused with the area of influence over the depth of the borehole in which gas is assumed to migrate and enter the headspace. The area of emission and the area of influence are not necessarily the same but are often misinterpreted by designers of gas protection systems.

The Pecksen equation states that:

surface emission rate of gas (l/h/m^2) = gas concentration × borehole flow rate (l/h)/10

An example of its use is given in Box 4.6.

Box 4.6 Example using the Pecksen method

Borehole flow rate recorded as 1 l/h
Methane concentration in borehole = 23%

Using the method proposed by Pecksen (1991)
Surface emission rate of methane = 1 × 0.23/10
Surface emission rate = 0.023 l/h/m^2

Flux box testing has shown that the Pecksen method is generally conservative by a factor of at least 10 and possibly up to 100 (Wilson *et al.*, 2004). Although it is widely used, its main drawback is that it neither takes account of the per-meability of the soil, nor of the depth of the headspace in the well from which the gas is sampled. For example the emissions from a source at depth overlain by an impermeable clay layer will be very different to those from a shallow source with a permeable cover, even though monitoring in boreholes may give similar readings for the gas concentration and flow rate. This must be taken into account when assessing the risk of gas surface emissions.

There is great uncertainty regarding the radius of influence around boreholes. Historically, recommendations were based on flow velocity readings taken in the open pipe. Most flow readings are now measured via a 4 mm diameter valve. However, flux box testing and comparison of the results with flow readings from both 19 mm standpipes and 50 mm standpipes with gas taps suggests that the relationship should always overestimate the actual surface emissions (as intended by Pecksen), often by a factor of up to 100.

A more robust method of estimating surface emission rates uses the measured differential pressure in boreholes. Most gas monitoring instruments in use today are capable of recording this parameter. The differential pressure in the well should also be less variable as it should depend less on the radius of influence. It will, however, be influenced by factors such as changing groundwater levels which trap and pressurise air in the top of well. There will also be an effect due to the lag between changing atmospheric pressure and the resultant changes in soil pore pressures.

The advantage of using this approach is that it is similar to other approaches used to model pressure driven flow of gases and vapours through the ground (for example the Johnson and Ettinger model).

The surface emission rate can be estimated as follows:

total gas flow from ground $Q_v = \left[\dfrac{K_i \gamma A i}{\mu}\right] \times$ methane concentration

Where:

K_i = intrinsic permeability of soil (m²) = $K_{Darcy} \times 1.02 \times 10^{-7}$

Q_v = flow of ground gas from the ground (m³/s)

γ = unit weight of landfill gas (N/m³)

μ = viscosity of gas being considered (Ns/m²)

A = surface area of migration (m²)

i = pressure gradient along migration route from ground as a fluid gradient = (gas pressure/unit weight of gas)/length of migration pathway (Pa)

The value of the permeability of the soil can be amended to allow for varying degrees of saturation of the soil. An example using this approach is shown in Box 4.7.

Box 4.7 Example of estimating surface emission rate using Darcy's law

Estimate the surface emission rate for a site with the following parameters:

Gas is migrating from a source at 4 m depth and is flowing vertically with a measured driving pressure of 50 Pa. The average methane concentration is 26%.

$K_{Darcy} = 5 \times 10^{-4}$ m/s

K_i = intrinsic permeability of soil in m² = $K_{Darcy} \times 1.02 \times 10^{-7} = 5 \times 10^{-4}$ m/s $\times 1.02 \times 10^{-7} = 5.1 \times 10^{-11}$ m²

γ = unit weight of landfill gas in N/m³ = 7.2

μ = viscosity of gas being considered in Ns/m² = 1.03×10^{-5} (for methane in this case)

A = surface area of migration in m² = 1 m² for a unit area

i = pressure gradient along migration route from ground = (50 Pa/7.2 N/m³)/4 m = 1.7

Total gas flow from ground $Q_v = \left[\dfrac{K_i \gamma A i}{\mu}\right] \times$ methane concentration

$Q_v = \left[\dfrac{5.1 \times 10^{-11} \times 7.2 \times 1 \times 1.7}{1.03 \times 10^{-5}}\right] \times 0.26 = 6.1 \times 10^{-5}$ m³/s/m²

The equations for diffusive flow that are discussed earlier in this section can also be used to estimate the vertical flow and surface emissions of gas due to diffusion.

SUMMARY: Gas and vapour generation and migration

The bulk gases (methane and carbon dioxide) are generated by the biodegradation of organic material. This can occur in any soils where organic material is present and especially in landfill waste. It also occurs where there are hydrocarbons in the ground. Most natural methane and carbon dioxide in peat and alluvium has already been generated and is largely trapped in the soil. Methane in coal is only released in significant volumes when a void is made in the coal (e.g. workings or a borehole). The methane then desorbs from the coal into the void.

Degradation of organic matter to generate ground gas is a complex biochemical and microbiological process. It is affected by variables such as: volume of degradable material, moisture content, nature of material, pH of soil or waste etc. which in turn affect the level of risk associated with a source. There are various methods that allow a simple estimation of methane and carbon dioxide generation, although on development sites it is rare that sufficient data is available to allow an accurate prediction of gas generation.

Gases other than methane and carbon dioxide can also be generated from non-biodegradable (commonly referred to as inert) wastes by other chemical processes in the ground. These gases include: hydrogen sulphide, hydrogen cyanide, carbon monoxide, ammonia and hydrogen. Vapours can be produced in landfills but probably the most common source of vapours in the ground is the presence of hydrocarbon contamination from fuel and chemical spills or leaks. In these cases the vapours are present as a result of volatilisation from contamination of the soil or groundwater.

The extent and rate of gas migration out of the ground or along in ground pathways is dependent on the rate of generation together with the presence of a migration pathway. Thus the geological conditions in and around a site have a major influence on the risk of gas migration. In order for sustained migration to occur, the gas must be replenished at source. Thus significant gas generation is usually required for large volumes of gas to migrate.

The first step in modelling gas migration (either horizontal or vertical) is to develop a conceptual model and identify the potential migration pathways. A simple mathematical assessment of gas migration can be achieved using Darcy's law to model pressure driven flow and Fick's law to model diffusion. These equations can model the migration of gases or vapours both horizontally and vertically in the ground and to estimate surface emissions.

More detailed modelling of gas migration can be undertaken using finite element packages, although the extent of the data required limits their usefulness. Conversely, a very crude estimate of the vertical migration of gas from the ground can be obtained from the measured borehole flow rates using an empirical relationship proposed by Pecksen (1986).

CHAPTER FIVE

Site investigation and monitoring

5.1 Strategy

In common with investigations of contaminated land the strategy for any site investigation and monitoring related to ground gas or vapours is to identify (or rule out) potential source–pathway–receptor pollution linkages and to develop a conceptual risk model for the site. This strategy is explained in detail in CLR 11 (DEFRA and Environment Agency, 2004). Ground gas is a contaminant and the identification of potential pollution linkages is no different from that of any other soil or groundwater contaminant.

Investigation and monitoring is a data collection exercise on the basis of which a conceptual site model can be established and potential pollutant linkages identified. All investigations should be phased: beginning with a simple gathering of existing information on the site (desk study). Subsequent phases are more targeted to examine issues identified from previous phases. In this phased approach the risk of not identifying key site hazards and pathways can be minimised.

Specific guidance relating to the investigation and monitoring of ground gas and vapours is described below.

5.2 General guidance

There is a wealth of information on the methods of data collection that may be appropriate for ground gas or vapour investigations:

- CIRIA, Report C659/665 Assessing risks posed by hazardous ground gases for buildings (Wilson *et al.*, 2006/2007)
- British Standards Institution, BS 10175:2001 Investigation of potentially contaminated sites – Code of Practice (BSI, 2001)

- NHBC, Guidance on evaluation of development proposals on sites where methane and carbon dioxide are present (Boyle and Witherington, 2007)
- CIRIA, Report 131, The measurement of methane and other gases from the ground (Crowhurst and Manchester, 1993)
- CIRIA, Report 151, Interpreting measurements of gas in the ground (Harries *et al.*, 1995)
- Institute of Waste Management, Monitoring of landfill gas, 2nd edn. (IWM, 1998)
- Environment Agency, Guidance for monitoring trace components in landfill gas (Environment Agency, 2004d)
- British Standards Institution, BS 10381: 2003 Soil quality – sampling, Part 7, sampling of soil gas (BSI, 2003)
- British Standards Institution, BS 8485: 2007 Code of practice for the characterization and remediation of ground gas in brownfield developments (BSI, 2007)

5.2.1 Desk study

Comprehensive risk management of ground gas or vapours must begin with a good desk study (see CLR11 and BS 10175) that is used to identify all potential sources of gas or vapour below, or in the vicinity of, a site and the likely potential for gas generation. This data is used to carry out a qualitative assessment of the risk posed by the presence of any gas and to design an appropriate ground investigation and gas monitoring regime. In addition, if assessing landfill emissions or investigating hydrocarbon contamination, the presence of trace gases or vapours may need to be considered, depending on the nature of the deposited waste or contamination.

5.2.2 Soil descriptions

Soil descriptions in accordance with BS 5930 (BSI, 1999) together with an estimate of degradable organic carbon (DOC) are absolutely vital to allow an assessment of future gas generation potential and to assess potential migration pathways. Estimating the DOC content of a material is, however, not straightforward. It is useful to divide samples taken during an investigation and weigh the various different fractions (soil, wood, cloth, leather, vegetable matter etc.). General descriptions of 'domestic refuse' or 'fill' are of limited use when trying to assess the risk of a gassing source, unless they are complemented by a more complete description of the proportion of the various constituent materials. This is particularly important where Part 2A assessments are being carried out.

5.2.3 Gas generation tests

The biological methane potential (BMP) is probably the best test to measure the gas generation potential from a degradable carbon-based material. The BMP test can take many months to perform and is not widely conducted by commercial laboratories. It can therefore be costly. An alternative procedure is to estimate aerobic degradation using the dynamic respiration index (DR4) test and the correlation with the BMP test given in Section 5.7. The DOC can also be estimated from the total organic carbon (TOC) determination, the latter test being a relatively cheap and quick to perform. The TOC concentrations can be correlated with known analytical conversion factors as defined by Hesse (1971) to determine the DOC content (see Section 5.7). Determination of the gas generation potential should be considered a routine requirement for Part 2A assessments.

5.2.4 Gas monitoring

Gas monitoring data must be obtained over a sufficient period of time and with a suitable number of visits over a range of atmospheric pressures. Other important factors to be considered are: the number and position of monitoring locations, response zones and the type of gas and other parameters tested for. All this data should be included in reports. The gas monitoring regime should be agreed with the environmental and development regulators at an early stage. There should be some flexibility in any monitoring programme to allow for variations if the early readings demonstrate that this is necessary.

The risks posed by the presence of gas cannot be assessed by looking at gas monitoring results in isolation from other data. A comprehensive desk study and site investigation are required to identify all the likely sources of gas below a site and their generation potential. This will allow the identification of potential sources such as mineworkings (these are particularly difficult to assess based on gas monitoring results as very large volumes can be emitted sporadically via discrete pathways, thus the gas emissions tend to be sporadic and spatially variable).

If the source of the gas is unknown then the risk cannot be assessed. A low concentration of gas that is caused by migration from an adjacent domestic landfill is of much more concern than a high concentration that occurs from a limited source such as occasional fragments of wood in fill material or pockets of peat. This is very difficult to identify from gas monitoring without supporting desk study information. Anomalies in the gas monitoring data should be investigated further (e.g. where gas is detected but there is no apparent source).

The quality of information and the overall degree of confidence associated with the analysis of that information should be sufficient to give a suitably robust basis for decision making. Where statistical analysis is carried out then the data set must be sufficient to allow this (e.g. statistics should not be used when there are only three sets of results). It is also important to submit the source data to environmental regulators along with a summary or graph of the data.

It is extremely difficult to zone a site into areas that have different levels of risk from ground gas, based on gas monitoring alone, unless the results can be correlated to some other clearly defined parameter (e.g. if gas is only encountered where the wells are installed in a particular soil type, and that soil type is not present over the whole site). On most sites there is generally insufficient data to allow zoning where the gas source is present across the whole site (e.g. if made ground is present across a site and gas is present in only one area with no correlation to the ground conditions, then it is invalid to zone, because the nature of gas monitoring does not give sufficient certainty to do this).

5.3 Measuring ground gas

A common oversight in gas monitoring is the failure to appreciate what is actually being measured. In practice the gas sample being measured is from the headspace within the standpipe where gas from the surrounding ground is allowed to accumulate. It is important to appreciate that it is not a true measure of the gas regime that actually exists in the soil pore space or soil matrix, which might be radically different at the soil microfabric scale. Nevertheless, measurement of the accumulated soil gas in a headspace is widely accepted as best practice in the UK. Indeed in many respects it can be considered to simulate the risk of gas ingress into a building from a gas-filled void in the ground and is therefore a valid test. However, there are a number of factors that affect the measurement of the gas in the headspace. They should be considered when interpretating the ground gassing regime and assessing the risks.

5.3.1 Headspace

When the borehole or probehole is initially constructed and the gas monitoring standpipe is sealed the air composition within the headspace will be in equilibrium with atmospheric air. Thus the concentration of gases such as methane and carbon dioxide will reflect the normal background concentrations in the atmosphere, typically less than 0.1% by volume in air. With time, however, ground gas from the soil pores will diffuse under concentration gradient or flow under pressure gradient into the headspace. A high concentration of ground gas in the soil pores will mix with atmospheric air in the headspace and will be diluted. Thus, at any time, the gas concentration in the headspace is a function of the volume of the headspace, the gas concentration gradient, the gas pressure gradient, and the gas permeability of the soil.

For soils of relatively high permeability, such as sandy or gravelly soil, equilibrium with the head space will be achieved relatively quickly. In clay soils of low permeability the time period will be considerably longer. The gas monitoring instrument records the gas regime in the headspace in terms of concentration and flow or pressure. Most modern ground gas monitoring instruments sample the gas by pumping a volume of gas from the headspace to the instrument. The sampling action changes the gas regime in the headspace. If the headspace volume is small compared to the volume sampling rate then the gas regime will be quickly depleted. Depending on how fast gas in the surrounding soil can replenish the headspace, the monitoring results can potentially significantly alter the gas regime, particularly in clayey soils of low permeability. For this reason it is good practice to take both an instantaneous reading of concentration and flow (or pressure) and steady-state readings and the time to reach steady state. These parameters are important when making any evaluation in terms of risk.

The head space will be affected by the groundwater levels. This can affect the response zone of the well to ground gas, although the effects are not clearly understood at present. Changes in groundwater levels can, however, cause suction or pressure within the head space if the groundwater level is above the top of the slotted section of standpipe.

5.3.2 Soil pore structure

The presence of the groundwater table or leachate and the degree of saturation of the soil can alter the gas transmission characteristics through the soil pore structure towards a borehole or gas monitoring standpipe. Wheeler (1988) has described how ground gas can move more freely through ground that is partially saturated than in situations where voids are completely water filled. The soil pore structure and the solid/water/gas phases can combine to produce porous systems with very different gas transmission properties. In saturated ground the ability of gas bubbles to move through the soil matrix from pore space to pore space requires the displacement of water from an adjoining pore space to accommodate the gas bubble. The ease with which the water/fluid can flow from one pore space to the next is a function of the hydraulic conductivity of the soil. In low permeability clayey soils, such as alluvium deposits, high gas pressures may be required to move water held in soil pores. If the gas pressure is too low, any gas may remain trapped in the pore space. In this way many organic estuarine alluvial soils can contain large quantities of ground gas which is only released when the soil structure is disturbed such as during the construction of a monitoring well (or the construction of piled foundations/ground improvement works).

5.3.3 Soil permeability

When deposited the vertical geostatic stress in a soil is higher than the lateral stress due to the self-weight of the overlying soil. The vertical compression is therefore greater than the horizontal compression. This results in the general property that the horizontal permeability can be up to several orders of magnitude higher than the vertical permeability. In man-made as well as natural soil the horizontal permeability tends to be larger than the vertical permeability (Lambe and Whitman, 1969). In compacted made ground, widely found on urban sites and a common source of ground gas, ratios of horizontal permeability to vertical permeability tend to be even larger than those in normally consolidated natural soils.

For this reason water in the soil matrix will flow more easily in a horizontal direction than vertically. Thus gas bubbles will tend to form and coalesce in a horizontal plane as water in the soil matrix is displaced in the preferential horizontal direction. As more bubbles accumulate in the horizontal plane they link together, forming networks which result in open cracks or fissures within the soil matrix through which gas can flow with little resistance (see Figure 5.1). Gas migrates through these open cracks or fissures either by gas generation pressure (advective flow), diffusion or buoyancy and escapes to atmosphere via the gas well installation (or other preferential vertical route).

Borehole

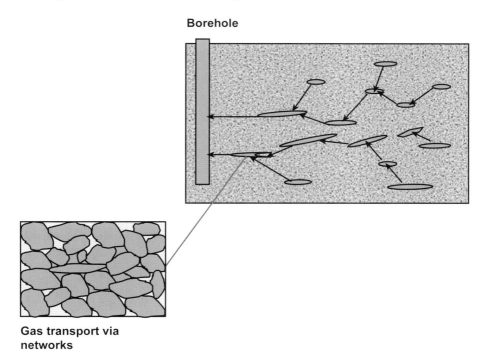

Gas transport via networks

Figure 5.1 *Gas flow in soil fissures*

5.3.4 Biological reactions

Changes that occur in the soil zone surrounding the borehole due its construction can introduce atmospheric air into this zone. This can result in aerobic methane oxidation. Various types of methane oxidising bacteria have been identified which are known as type 1 or type 2 methyotrophs that exist in aerobic soil zones. Thus it is not uncommon that methane readings in the borehole headspace may reduce with time as the surrounding zone of methane in the soil is oxidised to carbon dioxide and water vapour. Once the oxygen-rich atmosphere is consumed by the methylotroph bacteria and returns to anaerobic conditions then methane-producing soil bacteria will once again dominate and methane concentrations in the headspace will increase. This is more fully described in CIRIA Reports 130 and 151 and is a good reason to undertake longer term monitoring on some sites, and why monitoring in a newly constructed well is not always representative of the longer term gas regime below a site.

If wells are also used for water monitoring this can result in the borehole air being diluted with atmospheric air during each monitoring event. Hence the aerobic/anaerobic changes may be cyclic in these circumstances.

5.3.5 Implications for monitoring

It is extremely important, therefore, that sufficient gas monitoring readings are taken over a period of time and at an appropriate frequency between readings to allow the headspace to attain equilibrium with the surrounding soil gas pores. In sandy or gravelly soils of relatively high permeability equilibrium may be achieved in a matter of days after each reading and resealing of the standpipe. In contrast, in clayey or silty soils (greater than say 5% clay/silt content) of relatively low permeability it may take many weeks for the headspace to reach a state of equilibrium with the surrounding gas concentration in the soil pores. Indeed, it may never reach equilibrium as it is in a constant state of flux due to continuous changes in atmospheric pressure and wind velocity at ground surface which create a dynamic pressure gradient between the gas held in the soil pores and the gas in the headspace. In this respect it is questionable whether for most soils with greater than 5% silt or clay content (i.e. the majority of soils in the UK) the most onerous set of gas data can actually be recorded under falling atmospheric conditions that may only occur for a number of hours. Until further research is undertaken it would be wise to continue with the current practice of trying to obtain readings at low and falling pressure, especially where monitoring wells are located in a granular or open migration pathway that is outside a sealed gas source such as a capped landfill site.

At first sight it may seem that the true gas regime in the ground is not being recorded. In reality the borehole headspace measurement is a surrogate test of what actually happens and is a realistic model of how ground gas is likely to migrate through cracks or fissures in the soil (including flow paths created by construction activities, such as deep foundations, drainage or service infrastructure) and accumulate in a confined space within a building to give rise to a risk. There is therefore good justification for adopting the headspace measurement rather than attempting to determine the true gas regime within the soil pores. All guidance documents on the method of gas monitoring and interpretation of the soil gas regime rely on the headspace measurement and advocate monitoring under falling atmospheric pressure conditions to determine the most onerous concentration and borehole flow conditions.

5.3.6 Gas monitoring instruments

Detailed information on various gas monitoring instruments has been provided in the references cited in Section 5.2. The most common instruments are infra-red analysers. It should be noted that these instruments can be affected by the presence of hydrocarbons and can give unrealistically high values for methane (note that in such cases methane may also be present as a result of hydrocarbon degradation). Where hydrocarbon vapours are suspected of interfering with the meter it is good practice to take samples of gas for laboratory testing.

There are other cases where the instruments can give misleading readings (some instruments can interpret concentrations of hydrogen or ethane as carbon monoxide). Where unusual readings are obtained it is best to consult the manufacturer of the instrument to check if the results are due to the interference of the meter.

Where nitrogen is being recorded by the instrument it is wise to consult the manufacturer to ensure that the instrument is actually measuring the concentration of nitrogen in the ground. In the past some instruments have not actually measured nitrogen, but have calculated a value by subtraction of the methane, carbon dioxide and oxygen readings. The same approach has also been used by some laboratories when analysing gas samples. If there are other gases or vapours present this method will give erroneous and misleading results.

5.4 Measuring vapours

The assessment of risks from vapours requires the site to be comprehensively characterised in terms of the contamination below the site, because the most common models used in the risk assessment process (e.g. Johnson and Ettinger) are based on modelling the phase partitioning of the contaminant between the soil or aqueous phase and the vapour phase. Vapour monitoring results are not normally used directly in the risk assessment. The results can, however, be used to help validate the results of a risk assessment.

Monitoring for vapours is carried out in the headspace of the boreholes. Many of the comments regarding ground gas also apply to vapours. Vapours are usually monitored using photo-ionisation detectors (PIDs). Guidance on these has been provided in CIRIA Report C659/665. Sampling should follow the guidance in British Standard BS 10381:2003. Identification of specific vapours can be made using portable gas chromatographs attached to either a PID or flame ionisation detector (FID).

Monitoring of vapours inside buildings can also be carried out using sorption tubes that are fixed to walls and left in situ for a period of time before being removed and taken to a laboratory for testing.

5.5 Monitoring wells

The following guidance is based on the information provided in Wilson and Haines (2005) that was subsequently incorporated into CIRIA Report C659/665 (Wilson *et al.*, 2006/2007). It focuses on ground gas rather than vapours. For vapours the number of wells required to characterise a source will be driven by the extent of the chemical contamination of the ground or soil.

5.5.1 How many?

The number and spacing of monitoring wells required for any site should be based on informed judgement and the need to provide robust data for assessment and design. The location and number of wells are site-specific but there are some general rules that can be used to help provide a consistent approach. CIRIA Report 150 provides guidance on the positioning of wells and selection of appropriate grid patterns and should be essential reading for anyone planning or reviewing the results of a ground gas investigation. The guidance indicates that that following areas of a site should be targeted for well installation:

- Critical areas of the site where the desk study has identified a higher risk of gas being present, for example deeper areas of landfill, the perimeter nearest an adjacent landfill site, between the gas source and a receptor, or within zones of permeable geology that could provide migration pathways
- Areas of developments (or proposed developments) that are more sensitive to gas risk, for example below buildings, service pathways etc.

The spacing of wells is dependent on the location and number of potential gas or vapour sources and likely receptors. The permeability of the ground is also important as this will affect the radius of influence of the well. In less permeable ground more wells may be required.

Where there are no specific areas to target monitoring points, for example where the source is a consistent stratum below a site, then the most suitable method of setting out gas wells to give a representative indication of the gas regime is to use a uniform grid pattern (Raybould *et al.*, 1995). The spacing of the wells should vary according to the specific site conditions and the magnitude of the risk associated with the gas source. It is important to recognise that the purpose of collecting gas data is to allow an assessment of risk and provide design data for gas protection measures. It is not a research exercise.

Based on the guidance in CIRIA Report 150 a decision framework such as the one in Table 5.1 can be developed. However, three wells and four sets of readings should be considered an absolute minimum for even the smallest site.

Table 5.1 *Spacing of gas monitoring wells for development sites*

Gas hazard	Typical examples	Sensitivity of end use	Initial nominal spacing of gas monitoring wells[1-3]
High	On or adjacent to domestic landfills (see note 4), shallow mineworkings or shafts	High[4]	Very close
		Moderate	Close
		Low	Close
Moderate	Older domestic landfills (dilute and disperse, closed), disused shallow mineworkings, made ground with high degradable content, contamination by volatile organic compounds	High	Close/very close
		Moderate	Close
		Low	Close/wide
Low	Made ground with limited degradable material, organic clays of limited thickness, foundry sands	High	Close
		Moderate	Wide
		Low	Wide/very wide

[1] Very close = 0–25 m, close = 25–50 m, wide = 50–75 m, very wide = 75–100 m
[2] The initial spacing may need to be reduced if anomalous results indicate that this is necessary to give a robust indication of the gas regime below a site. To prove the absence of gas closer spacing may be required
[3] The spacing assumes relatively uniform ground conditions and the gas source present below a site. The spacing will need to be reduced if ground conditions are varied or if the investigation is trying to assess migration patterns from off site
[4] Placing high sensitivity end use on a high gas hazard site is not normally acceptable unless source is removed or treated to reduce gassing potential

High sensitivity end uses are developments such as housing with individual gardens or schools. An example of a moderate sensitivity end use is an office development. A warehouse is a typical low sensitivity end use.

A network of wells which is designed merely to investigate whether or not gas is present (detection monitoring) may be less extensive than one installed to determine the rate and extent of gas migration (assessment monitoring) or to monitor remedial activities. The following factors also need to be considered when using the table:

- For a relatively consistent, low generation potential source, a few widely spaced wells may be adequate to characterise the site
- Where zoning of gas protection measures is proposed on a site, the number of wells required increases
- When assessing migration from an off-site source, a significant number of wells may be required, particularly to demonstrate that gas migration is not occurring. The guidance from Waste Management Paper 27 (Department of Environment, 1989) is still applicable in this respect (see Table 5.2)

Table 5.2 *Monitoring well spacing to detect off-site gas migration*

Site description	Monitoring borehole spacing (m) Typical range
Uniform low permeability strata (for example clay); no development within 250 m	50–150
Uniform low permeability strata (for example clay); development within 250 m	20–50
Uniform low permeability strata (for example clay); development within 150 m	10–50
Uniform matrix dominated permeable strata (for example porous sandstone); no development within 250 m	20–50
Uniform matrix dominated permeable strata (for example porous sandstone); development within 250 m	10–50
Uniform matrix dominated permeable strata (for example porous sandstone); development within 150 m	10–20
Fissure or fracture flow dominated permeable strata (for example blocky sandstone or igneous rock); no development within 250 m	20–50
Fissure or fracture flow dominated permeable strata (for example blocky sandstone or igneous rock); development within 250 m	10–50
Fissure or fracture flow dominated permeable strata (for example blocky sandstone or igneous rock); development within 150 m	5–20

Some examples of monitoring well spacing are shown in Figure 5.2 (colour section).

5.5.2 Gas spike surveys and wells in trial pits

Gas spike surveys use an iron bar driven into the ground by a sledgehammer to form an open hole. Gas monitoring is then carried out in the open hole. There is no excuse for using shallow spike surveys in a modern investigation. Because they are open they are unreliable and do not provide any useful data on which to base a ground gas assessment. This is because the absence of gas is never conclusive and the concentrations of gas identified are not a robust indicator of the gas regime in the ground (because of dilution by air in the hole and the limited depth of penetration, typically a maximum of 1 m). Regulators should discount the results from spike surveys. There are more reliable methods of installing gas wells that are quicker and easier (and therefore cheaper).

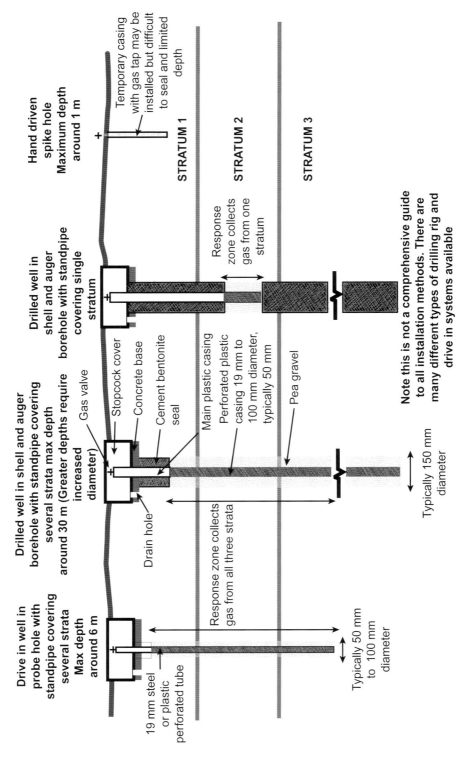

Figure 5.3 *Typical gas well installations*

Another technique which is still used is to install gas monitoring wells in trial pits. This can be done but the well installation should be constructed in accordance with the guidance in CIRIA Report 131, with a robust polythene sheet placed over the top of the pit and sealed down the edges and to the standpipe, to prevent escape of gas through the disturbed backfill. It is also necessary to leave the installation longer (at least one month) before reliable readings can be obtained, and more readings are required to be sure of the results, again because of the ground disturbance. As a result the use of drive in gas wells is usually a more cost-effective and technically sound way of installing wells alongside trial pits.

5.5.3 What diameter monitoring well?

The two most common methods of installing gas wells are either in shell and auger or rotary drilled boreholes of 150 mm diameter or greater, or in window sampler holes with either 50 mm or 19 mm diameter standpipes installed (see Figure 5.3).

Experience on many hundreds of sites suggests that there is no practical difference between measured gas concentrations and flow rates obtained from 19 mm drive in standpipes and 50 mm wells installed in boreholes (even though theory may suggest otherwise). Typical gas monitoring results from a development site in Wales are shown in Figure 5.4. The wells were all installed in the same made ground to similar depths. The made ground comprised a mixture of ash, coal gravel, burnt colliery shale, brick rubble and clay with mudstone cobbles and boulders. It was generally inert with occasional wood fragments, peaty material and organic clays. There is no obvious correlation between methane flow (i.e. methane concentration × borehole flow rate) and the diameter of the standpipe.

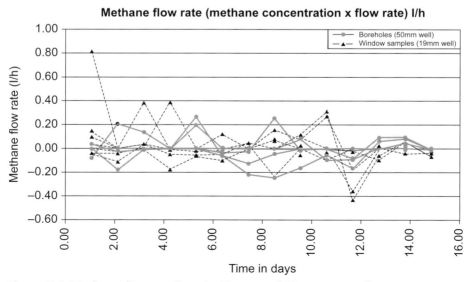

Figure 5.4 *Methane flow readings in 19 mm and 50 mm gas wells*

Both methods are widely used when investigating ground gas problems. The absence of any correlation is probably because the wells are normally left closed and do not vent continuously. Therefore the well is simply a space that comes into equilibrium with the ground gas regime. In addition, if there are any variations due to the well diameter they are likely to be masked by a multitude of other variables in ground gas regimes, such as complex responses to atmospheric pressure variations and the oxidation of methane.

For similar reasons to those above it is also considered that the well diameter will have no significant effect on the results of vapour monitoring.

5.5.4 How deep should the monitoring wells be?

The depth of gas wells is also based on a characterisation of geological and hydrogeological conditions at the site and on the perceived level of risk associated with ground gas or vapours. The depth of wells should be sufficient to intercept any gas or vapour sources or migration pathways. The present authors have seen several examples where gas has not been encountered but where the wells have not been installed deep enough to intercept the source. Examples include:

- A site investigation report was submitted to a local authority for approval, for a site overlaying an historic sea inlet with alluvium present. Boreholes were installed, encountering thick peat at several metres depth, separated from the shallow made ground by a clay layer with gravel lenses. The response zones for the monitoring wells were in the shallow made ground and did not intercept the peat. The developer argued that migration of gas to the proposed development was prevented by the impermeable clay layer. However, there was no gas data from the peat layer to support this conclusion. The regulator determined that no account had been taken of the recommended foundation construction of piles extending through the peat, or of gravel lenses offering potential migration routes. The development was delayed whilst additional wells with appropriate response zones were installed and monitored. Monitoring initially encountered relatively high concentrations of methane, and gas protection measures were installed

- A site investigation report was submitted containing gas-monitoring information obtained by a shallow spike survey. The report argued that as only low concentrations of methane (<1%) had been identified during the spike survey, this indicated a low gas regime on the site. However, the spike survey was at shallow depth and had not penetrated the main body of gassing material below the site. The local authority determined that (referring to CIRIA guidance), positive gas results encountered during a spike survey are generally representative of higher gas concentrations being present within the ground, and cannot be considered to give an indication of actual gas concentrations or generation rates, or be used on their own for gas investigations for proposed development sites. The development was delayed whilst gas wells were installed and monitored. Monitoring results indicated that higher levels of gas were present and that basic gas protection measures were required

- A site investigation was undertaken for a development on a sloping site at the edge of a river flood plain. The main part of the development was to be located over peat deposits and the higher ground was underlain by glacial till. Gas monitoring wells were only installed on the higher ground within the glacial till and gas was not encountered. During construction, concerns were raised about ground gas when bubbles were observed in standing water in pile cap excavations over the area of peat. Gas was suspected within the peat and gas monitoring wells were installed. These encountered elevated levels of methane up to 90% v/v and gas protection measures were required

If there are different potential sources or discrete pathways a number of wells may need to have their response zones sealed into different strata rather than the common method of installing a gravel surround along the whole length. An example is given in Figure 5.5 where there were two potential sources of gas. One was the deep sand layer where migration from the landfill (on the left-hand side) occurred and the other is the shallow made ground to the right of the site. Wells were installed and sealed into the different strata to determine the ground gas regime from both sources. If the deep wells had not been installed then it is likely that the extent of gas migration below parts of the site would not have been identified. In addition some wells were installed above the sand layer to identify if any significant vertical migration was occurring.

Installing wells that have response zones in more than one source or pathway is bad practice and should be avoided as it usually makes it very difficult to interpret the gas monitoring results (it is virtually impossible to identify where the gas is coming from).

5.6 Monitoring protocols

5.6.1 What to monitor?

Considering that the main cost of gas monitoring is incurred in getting the person to site, then a few minutes extra time at each well collecting the necessary data is money well spent.

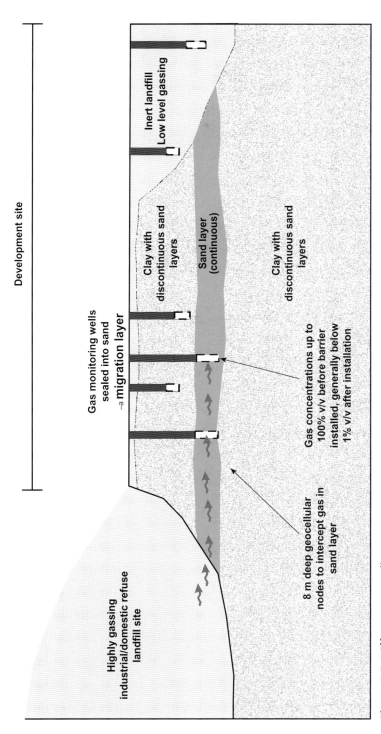

Figure 5.5 *Different gas well response zones*

It is important to record (in addition to the monitoring parameters from the instrument such as gas concentrations, borehole flow rates, atmospheric pressure and vapour concentrations where appropriate):

- Changes in atmospheric pressure over the two days preceding monitoring (this can be obtained from the Met Office (www.met-office.gov.uk))
- Condition of ground surface
- Any activities in the area that may affect readings (e.g. dewatering or excavation). In one case at a football stadium, operation of the underpitch air drying system was affecting the borehole flow rate
- If readings are varying and by what degree. For example does the gas concentration record a peak and then reduce and over what time?
- Any odours from the wells
- Groundwater levels
- Weather conditions
- Temperature within the well compared to ambient conditions (elevated temperatures within a well are an indication that significant biodegradation is occurring)
- For vapour surveys monitor and record if there is any free product in a well

5.6.2 How long to monitor?

The following guidance is specifically for monitoring ground gas. There is no guidance for vapours and each site will require a site-specific assessment to determine the appropriate amount of monitoring. There are no binding rules for the length of time to monitor as every site is different. However, there are some general rules that can used to guide decisions. The main concern is to collect sufficient data on which robust decisions can be made and the worst case gas conditions identified or estimated for a site. The two key factors in determining the amount of gas monitoring data are:

- Gas generation potential of source
- Sensitivity of end use

Guidance on the number of readings, based on these factors, is provided in Table 5.3 (see colour section). In addition the variation of results and any anomalies will influence whether more than the suggested number of readings are required. It should be noted that the suggested periods and frequency of monitoring in Table 5.3 are unlikely to be applicable to monitoring in or around Part 2A sites.

It is important to consider the benefits of collecting any extra data and whether or not any useful purpose will be served. Ground gas investigations are usually intended to provide robust data for the design of protection. They are not normally designed as an academic or research exercise.

5.6.3 Examples of using Table 5.3

It is not always necessary to comply absolutely with Table 5.3 (see colour section). The following examples show how variations can be acceptable, provided they are justified.

Variation in number and period of monitoring

A small site was underlain by made ground and gas migration into a housing development (high sensitivity) was a concern. A comprehensive site investigation had determined that there was no significant biodegradable content in the made ground (very low generation). Gas monitoring had been carried out in 14 wells for periods of between three weeks and 12 months. Typically for a site of this size monitoring would only have been undertaken in three or four wells. Table 5.3 suggests a minimum of six readings over three months. In most of the wells the number of visits exceeded the six required. However, the period of monitoring was generally just less than three months for most wells with one significantly exceeding it. The overriding requirement was to obtain a reasonable spread of readings to be able to predict the worst case gas regime in the ground and allow the design of a reasonable gas management strategy. In this case given the excess number of readings and shorter monitoring periods, together with the number of wells that had been installed, it was considered that overall the number of gas readings achieved the aims of Table 5.3.

Removing a migration pathway

A small housing development (high sensitivity) was proposed adjacent to a landfill that was known to contain domestic refuse and was generating landfill gas. The landfill was completed in the 1970s (moderate generation). A shallow migration pathway was present (sand and gravel) that could potentially allow gas to migrate to the site.

Gas monitoring was carried out four times over a period of one month. In accordance with Table 5.3 at least 12 readings would be required over a six-month period and possibly more to characterise the risk of gas migration. However, the developer wanted to start work before this period could be achieved. Therefore an in ground gas barrier was installed to prevent gas migration. This removed the need to fulfil the requirements of Table 5.3.

A similar approach can be taken where shallow mineworkings are grouted up, thus effectively removing the source of gas (assuming there are no other sources).

Will additional readings change the characterisation?

An office development was underlain by made ground with a low degradable content that comprised pieces of wood and thin layers of topsoil (low generation potential). The development was considered to be moderately sensitive.

Table 5.3 suggests six sets of readings taken over a period of three months are required but only four sets of readings had been obtained. However, the maximum values obtained were:

methane – 12%, carbon dioxide – 7% and borehole flow rate 0.6 l/h

These results gave a gas screening value of 0.072 which indicated characteristic situation 2 (see Chapter 6). However, the results only just exceeded the threshold for this characteristic situation and given the nature of the source it was considered unlikely that further results would increase the characteristic situation (either the concentration or flow rate would need to increase by a factor of 10) and so further monitoring was not required.

5.7 Flux boxes or chambers

Flux chambers can be an important tool in assessing the risk of surface gas or vapour emission and migration into buildings. The technique involves placing a box over the ground surface and measuring the accumulation of gas inside the chamber over time (see Figure 5.6).

Flux chambers do, however, have limitations and in particular they should only be used in conjunction with borehole and other data to give an overall interpretation of gas emissions from a site. They should never be used to identify gas migration outside a gassing site, since the resulting surface emissions usually occur at discrete points in the ground and it is virtually impossible to locate flux chambers over a particular emission point. They can, however, be used to help quantify emissions from known zones of surface migration.

Figure 5.6 *Flux chamber testing*

If the gas source is at depth then the flux box will not identify the actual gas regime. If the development includes piled foundations these could introduce a new pathway to the surface and the flux box results may be invalid. Similarly if a highly gassing source is at depth in an infilled quarry, with overlying soils that are not generating gas, the flux box tests will be of limited value in estimating the likely risk of lateral migration at depth into the surrounding permeable strata (see Figure 5.7).

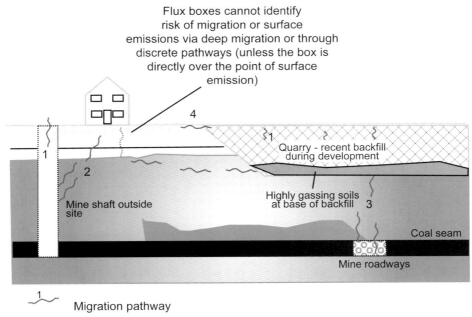

Flux boxes cannot identify
risk of migration or surface
emissions via deep migration or through
discrete pathways (unless the box is
directly over the point of surface
emission)

4

1

Quarry - recent backfill
during development

2

Highly gassing soils
at base of backfill   3

Mine shaft outside
site

Coal seam

Mine roadways

1
Migration pathway

Figure 5.7 *Inappropriate use of flux boxes*

5.8 Flame ionisation detectors (FID)

FIDs can detect flammable gases at very low concentrations (1 ppm) and are typically used to monitor surface emissions from the ground (as a surface sweep) or for internal monitoring inside buildings where migration of flammable ground gas is suspected. They can also be used in conjunction with a gas chromatograph where there is concern that hydrocarbon vapours are influencing the methane results on an infrared gas analyser.

5.9 Specialist testing

5.9.1 Laboratory testing of gas samples

Gas samples can be taken in sample containers and sent to a laboratory for analysis of the composition of the gas. The results can be compared to the in situ monitoring results. This is not routine practice, but is useful where anomalous readings have been obtained, or on more sensitive or difficult sites. Analysis for trace gas and vapours can help identify the origin of the ground gas.

The type of sampling container and method of sampling will depend on the gas or vapour being analysed and most laboratories will give advice on the most appropriate method. The most common method of bulk gas sampling (methane, carbon dioxide, carbon monoxide, hydrogen sulphide, nitrogen and oxygen) is by Gresham pump and tube.

A hand pump is used to pump gas into a cylindrical stainless steel or aluminium container. The standard Gresham tube, when filled to its maximum of 14 bar, will contain about 0.75 l of sample gas from the well. Care should be taken when using Gresham tubes to take gas samples from 19 mm diameter boreholes that gas is not drawn from the surrounding ground because the volume of gas in the headspace is low. The volume of the headspace for 19 mm and 50 mm diameter wells at various depths is given in Table 5.4.

Table 5.4 *Headspace volumes*

Depth of headspace in borehole (m)	Volume of gas (l)	
	19 mm diameter	**50 mm diameter**
1	0.28	1.96
2	0.57	3.92
3	0.85	5.88
4	1.13	7.84
5	1.45	9.80

If gas samples are taken from 19 mm standpipes with a headspace less than 3 m the samples may be drawing gas from the surrounding ground if the tube is filled to its maximum volume, and this should be considered when assessing any difference in results. Smaller size sampling tubes can also be used.

Adsorptive tubes are sometimes used to take samples for analysis for VOCs. A 4 mm diameter hard glass or stainless steel tube is filled with granular adsorbent (e.g. Tenax tubes). The air to be sampled is pumped through the tube using a small battery powered pump and the VOCs in the air are adsorbed and condensed. The tube is then sent to a laboratory for analysis.

5.9.2 Gas generation tests

The biological methane potential (BMP or BM100) is probably the best test to measure the gas generation potential from a degradable carbon-based material. The BMP test simulates the degradation of organic matter in the laboratory, typically over a period of up to 100 days, although it can take much longer. A known weight of sample is sealed in a glass vessel. The head space of the vessel is purged of oxygen and the volume of methane generated with time is measured by simply collecting the gas over water. The BMP test can take many months to perform and is not widely conducted by commercial laboratories. It can therefore be costly. An alternative procedure is the Fibre Cap Test (Kitcherside *et al.*, 2000) which is an accelerated BMP test.

The DR4 test provides a measure of the aerobic biodegradability of a sample over four days (Environment Agency, 2005d). The carbon dioxide production is measured as the sample degrades to give an indication of the biodegradability of the test material. This is converted to oxygen consumption units and the results are reported on both a dry matter and loss on ignition (LOI) basis (i.e. mg O_2/kg dry matter and mg O_2/kg LOI). The results can be used to estimate the BM100 results using the correlation in Figure 5.8, although it is preferable, if possible, to determine a site-specific correlation.

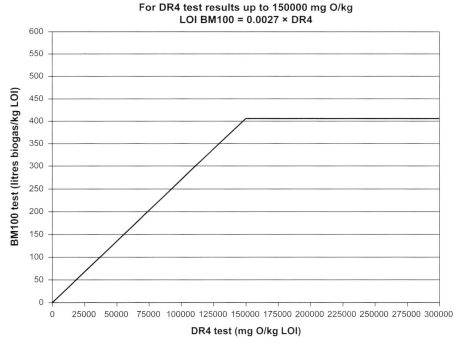

Figure 5.8 *Correlation between DR4 tests and BM 100 tests (Environment Agency, 2005d)*

These tests are particularly important for Part 2A assessments as they allow an estimate of the volume of gas likely to be generated. Understanding the volume of gas that will be produced is more important than the concentration of the gas when carrying out risk assessments.

5.9.3 Carbon content and other soil and groundwater tests

TOC gives a crude and conservative estimate of the amount of biodegradable material that is present in soil or waste. It includes readily available and non-readily available carbon. In addition, carbon is present soils in inorganic form such as natural carbonates, limestone and chalk but this is not part of the TOC. The degradable carbon content comprises both readily available carbon (e.g. cotton fibre, paper and wood products) that degrades relatively quickly and non-readily available carbon (e.g. plastics) that will break down and degrade very slowly.

The readily biodegradable fraction of the municipal waste, which can produce landfill gas, is primarily made up of cellulose and hemicellulose, although not all the cellulose in waste is available for biodegradation.

Therefore a more realistic estimate of gas generation potential can be achieved from the determination of the DOC content of the soil or waste. This can be estimated from a known analytical conversion factor of 1.33 as defined by Hesse (1971), i.e. $DOC = \frac{TOC}{1.33}$.

An alternative approach is to consider the cellulose plus hemicellulose to lignin ratio (Pueboobpaphan and Toshihiko, 2007). The use of cellulose plus hemicellulose to lignin ratio (C + H)/L is based on the premise that cellulose and hemicellulose are easily consumed in anaerobic biodegradation while lignin is not. The (C + H)/L ratio would have a larger value in the case of easily degradable organic matter such as food and paper, and a smaller value in the case of slowly biodegradable substances like wood and newsprint. A comparison of the (C + H)/L ratio and the gas generation from the BMP 100 test is shown in Figure 5.9.

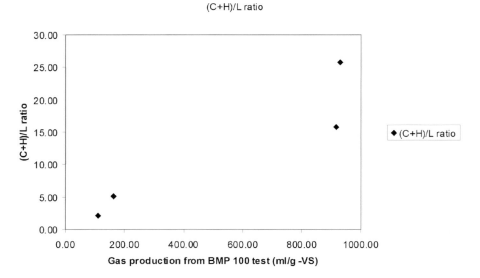

Figure 5.9 *Comparison of gas generation with (C + H)/L ratio*

The parameters required to determine the (C + H)/L ratio can be determined in a three stage test as follows:

1. Neutral detergent fibre (NDF), - neutral detergent fibre is a measure of total fibre in a sample, i.e. 100% of the hemi-cellulose, cellulose and lignin is recovered
2. Acid detergent fibre (ADF), dissolves hemicellulose leaving a residue of cellulose and lignin. The difference between tests 1 and 2 is therefore a measure of the hemicelluose content
3. Acid detergent lignin (ADL) tests determine the lignin and cellulose content of the residue from test 2. Cellulose is dissolved from the sample leaving a lignin residue. The difference between tests 2 and 3 represents the cellulose content. Subtraction of the cellulose content from the initial mass of residue in test 3 represents lignin

These tests can give a more accurate assessment of the amount of DOC in a sample and the results can be used to give an assessment of the volume and rate of landfill gas production, for example using GasSim (see Section 4.2). However, caution is needed when applying these tests to samples with a high soil or mineral content as they were originally designed to measure the fibre content of animal feed stuffs. They can give misleading results on samples with a high soil content because the reagents can dissolve minerals in the soil as well as degradable materials.

Similarly, to determine the nature and concentration of carbon in the landfill leachate, samples can be tested for biochemical oxygen demand (BOD) and chemical oxygen demand (COD), dissolved methane, DOC and volatile fatty acids (VFAs). If the BOD/COD ratio of leachate is greater than 0.4 this indicates that microbiological activity within the waste is still occurring and the population of bacteria is likely to increase giving rise to increased landfill gas production. In contrast if the BOD/COD ratio is less than 0.4, such as in leachate from old landfills, this indicates that microbial activity is declining and that gas generation is also declining (Crawford and Smith, 1985). This relationship does not always apply to very old landfills or very low generation sources that do not have elevated BOD and COD in the leachate.

Methane production can also be estimated from the COD value of leachate using the following formula:

$$\text{gas production} - \text{methane (m}^3/\text{tonne)} = 0.35\,\text{m}^3/\text{kg COD}$$

For this to apply to organic degradation in soil, waste or made ground a COD value for the complete anaerobically decomposed material will need to be determined, i.e. the residual COD. The potential for gas generation for a sample is then:

$$(\text{measured COD} - \text{residual COD}) \times 0.35\,\text{m}^3 = \text{biogas methane potential (m}^3/\text{t of material)}$$

Where the COD is measured in kg of oxygen/t of material.

5.9.4 Carbon dating

Carbon dating can be useful to identify the specific source of soil gas. In the case of methane carbon dating has been used to identify whether the source is from coal mines or mains gas. The method relies upon the ratio of carbon isotopes (carbon 12 to carbon 14) in the gas source. Methane derived from ancient geological sources (e.g. biogenic sources such as coal seams or petroleum) will have a 'finger print' in terms of the carbon isotope ratio from methane derived from more recent sources such as plants, animals etc. The method may not always work for methane from landfill sites as degradable waste may originate from both geologically ancient and recent organic sources. Further information can be found in CIRIA Report 131.

5.10 Future directions: continuous monitoring

Generation and migration of ground gas is a complex process that is difficult to monitor. The existing approach relies on discrete measurements of the gas concentration and borehole flow rate from which representative ground gas concentrations and gas migration potential are inferred. Because the data is generally poorly resolved spatially and temporally there can be large uncertainties in both the data and the inferences made from it.

Poor spatial resolution is a result of the physical inaccessibility of the subsurface which must be sampled by discrete boreholes. Where improvements in spatial resolution are required a greater number of boreholes is usually necessary and this will have a direct impact on the cost of an investigation. Temporal resolution may be improved as a result of recent technological advances in instrumentation without significantly increasing the cost of investigation, although at present it would require a substantial up-front investment in new monitoring equipment.

A new monitoring instrument known as the Gasclam has been developed by a joint venture between Salamander and the University of Manchester. It allows continuous unmanned data collection from a 50 mm diameter monitoring standpipe. It records methane, carbon dioxide, oxygen and hydrogen sulphide concentrations together with atmospheric pressure, borehole pressure, temperature and piezometric head (see Figure 5.10). The Gasclam unit meets the requirements for explosion protected electrical equipment.

Some currently available gas monitoring instruments can be set to data log readings at very close time intervals (e.g. gas data instruments) thus allowing some of the uncertainty in the data to be addressed. The present authors have used normal instruments in this manner, but they require special arrangements to prevent theft or vandalism if they are to be left for a lengthy period of time and the rechargeable batteries only have a limited lifespan. The Gasclam has overcome

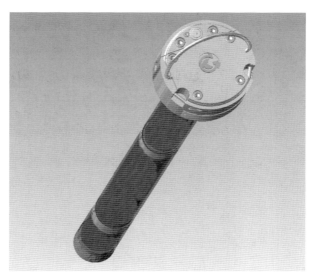

Figure 5.10 *Prototype of a Gasclam® monitoring unit*

these issues to some extent because it can fit inside a normal 50 mm diamter well without any modifications and can operate for a long period of time (although it would be wise to ensure that the well covers are extremely secure).

A number of benefits arise from the ability to continuously monitor ground gas concentrations in boreholes. It directly addresses the poor temporal resolution of the observations.

The gas concentration may vary more often than the sampling frequency, in which case measurements will not be representative, although the extent to which this is critical depends on the variation in the gas concentration. Periods of continuous monitoring can begin to quantify the variation in the gas concentration thereby allowing not only a gas concentration to be recorded but also a confidence limit on the concentration. The confidence limit can also be narrowed to a specified level by monitoring for a longer period. Ultimately, permanent use of a continuously monitoring device should return a representative gas concentration with no error due to temporal variation, althought is only likely to be practical on sites where very long-term monitoring is required (e.g. licensed landfill sites). The benefit of continuous monitoring in overcoming this type of mismatch in sampling frequency and variability in the gas concentration is clearly shown in data from a site which was thought to show a high gas concentration only at Christmas (see Figure 5.11).

The 'Christmas borehole' is located on the perimeter of a landfill. Monthly monitoring results indicated that gas migration problems only occurred at Christmas time. A period of continuous data collection has overcome the problem, which arose from the mismatch between the sampling frequency

Methane, carbon dioxide and oxygen

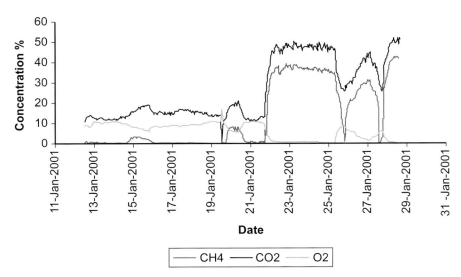

Figure 5.11 *Continuous gas concentration data from the 'Christmas borehole'*

(monthly) and the varying concentration. The continuous data clearly showed that the methane concentration was not only high at Christmas, although it was still variable. However, this phenomenon may also have been picked up by increasing the frequency of the conventional monitoring, to say daily events, in response to the sporadic spikes that occurred.

In addition to reducing the uncertainty in gas concentration measurement in a particular borehole, continuous monitoring should allow correlations of gas concentration and other environmental parameters, such as variations in atmospheric pressure, to be more meaningful.

Continuous monitoring using a device that controls the venting of the borehole can allow the gas production rates in the sampling borehole to be characterised. This can be done by monitoring the return to a starting gas concentration after the concentration has been decreased by pumping.

The main factor that is likely to limit the uptake of this technology in the short term is the cost of the units. The cost is comparable to a conventional monitor and therefore it is unlikley that many consultants will be able to afford to have these set up on more than one site at any one time. Given that ground investigation consultants can be monitoring several sites at any one time this is a significant constraint. Therefore their use needs to be integrated with a conventional monitoring programme and targeted at sites where there is likely to be most benefit from reducing uncertainty in the data. However, with time the unit cost should drop so that its use becomes increasingly viable on a greater range of sites.

SUMMARY: Site investigation and monitoring

The strategy for any site investigation and monitoring related to ground gas or vapours is to identify (or rule out) potential source–pathway–receptor pollution linkages and to refine the initial conceptual risk model for the site. There is a wealth of information on the methods of data collection that may be appropriate for ground gas or vapour investigations that is published by the Environment Agency, NHBC, British Standards and CIRIA.

Comprehensive risk management of ground gas or vapours must begin with a good desk study (see CLR11 and BS 10175) that is used to identify all potential sources of gas or vapour below, or in the vicinity of, a site and the likely potential for gas generation. The risks posed by the presence of gas cannot be assessed by looking at gas monitoring results in isolation from other data. The desk study and site investigation are required to identify all the likely sources of gas below a site and their generation potential. If the source of the gas is unknown then the risk cannot be assessed. The next stage is to carry out a site investigation to obtain good soil descriptions, carry out laboratory testing and undertake gas or vapour monitoring. Spike tests (gas readings in shallow open holes formed by driving a spike into the ground) are of no use in ground gas investigations.

Gas monitoring data must be obtained over a sufficiently long period of time and with a suitable number of visits over a range of atmospheric pressures. Other important factors to be considered are the number and position of monitoring locations, response zones and the type of gas and other parameters being tested for. The number of wells, their depth and the amount of monitoring will always be site specific but some general principles and guidance are provided to give a consistent framework for such decisions. All this data should be included in reports. The quality of the information and the overall degree of confidence associated with the analysis of that information should be sufficient give a suitably robust basis for decision making.

It is important to realise that, in practice, the gas sample being measured is from the headspace within the standpipe where gas from the surrounding ground is allowed to accumulate. This has implications for the analysis and assessment of data collected from wells.

Guidance on gas monitoring and what to record is given together with guidance on where to use techniques such as flux boxes and FIDs. Specialist testing that can be used to supplement gas monitoring includes: laboratory testing of gas samples, gas generation tests, carbon dating and tests to determine the degradable content of soils or waste.

This chapter concludes with a discussion of new methods of continuously monitoring gas wells over a period of time and the implications that this will have for gas risk assessment.

CHAPTER SIX

Assessment of results

6.1 General considerations

The analysis of data for a gassing site or site affected by gas migration should consider all the data together. One set of data should not be used in isolation to assess the possible risks (for example just looking at gas monitoring results). It is not possible to interpret gas monitoring results without a desk study and ground investigation that identifies possible gas sources and migration pathways.

The following discussions are intended to assist the reader in following the risk assessment process that is discussed in CIRIA Report C659/665 (Wilson *et al.*, 2007) and to follow the flow charts that are included in that document.

6.2 Conceptual model

The first step in any assessment of a site affected by ground gas or vapours is to construct a conceptual model. The desk study and site investigation data should be used to develop a conceptual model. An example is shown in Figure 6.1 and the model should be revisited and refined as and when more data becomes available. When designing the gas protection the model should be amended to include any gas protection that is deemed necessary.

The conceptual model should give a concise description of the ground conditions and gas model for a site. The best and most easily understood way of doing this is to draw a diagrammatical cross-section of the site (Figure 6.1). Another example is shown later in Figure 6.4.

The conceptual model should include:

- Ground conditions below a site
- All potential sources of ground gas or vapours

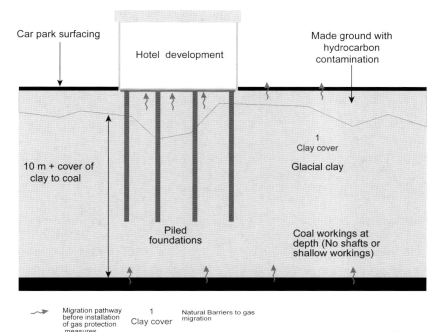

Figure 6.1 *Example of a conceptual model*

- All potential migration pathways
- All potential receptors
- Natural barriers to gas migration (e.g. clay soils surrounding a landfill site or overlying mineworkings as in Figure 6.1)

Further guidance on conceptual models is provided by the Environment Agency (2001) and in CIRIA Report C659/665. The Environment Agency report gives the following excellent advice

> The development of a conceptual model is emphasised and this must identify and consider all the relevant aspects of the flow system and the contaminant transport processes, including the source of contamination, the pathways of flow and transport and all of the potential receptors. Clear graphical presentation of the conceptual model is recommended.

> The conceptual model and the mathematical model should be continually challenged and updated throughout the modeling exercise.

> The development of a conceptual model will involve a number of assumptions regarding system behaviour. The assessment must take these assumptions and uncertainties into account and decide whether these are important, i.e. it may be acceptable to adopt a relatively simple mathematical model of contaminant transport, or alternatively our understanding and definition of the system behaviour may be so poor that development of a mathematical model is inappropriate, and the first priority should be to obtain further site-specific data.

6.3 Analysis of ground gas data

Guidance on the assessment of gas data is provided in CIRIA Report C659/665. The following paragraphs provide some useful advice on specific trends and patterns to look out for in gas monitoring data.

6.3.1 Outliers and anomalies

If sufficient results are available it is useful to identify any potential outliers in the data and determine if these indicate occasional peaks or could be representative of more persistent peaks (see Section 5.10). This is difficult with a low density data set.

Any results that do not fit with the conceptual model should be identified (anomalies) and justification provided that they are unrepresentative results or are indicative of the gas regime on site. For example, on one development site the only source of gas was the presence of small volumes of carbonate material in the natural glacial soils. Gas monitoring was available for several other sites in the area where there was no source of carbon dioxide other than the glacial soils and this indicated a general trend of carbon dioxide concentrations up to 3% volume which represented natural background concentrations. In the development site the gas monitoring had one single result of about 7% in one borehole (out of about 30 readings in five boreholes). In this case further monitoring was carried out and the 7% reading was shown to be unrepresentative of the sustained long-term gas regime below the whole site and was discounted in the assessment.

6.3.2 Graphs

Graphs are a useful tool for identifying trends in gas data. One common trend that has frequently been observed on sites where there is limited gas generation is for a peak in methane and carbon dioxide readings for the first month or so and then a gradual reduction to negligible levels (see Figure 6.2). This is thought to be as a result of disturbance of the soil caused by installation of the boreholes, although the precise mechanism is not fully understood. This effect is not seen at more highly gassing sites. Other trends that have been observed were variations in gas concentration with different surface conditions (e.g. higher gas concentrations observed in winter, when the surface is wet or frozen, than in summer when the ground is dry and gas can escape through the surface).

The graphs are used to highlight trends in the data, joining up the discrete data points with lines does not imply there is a direct linkage or linear relationship between each monitoring result. Because of the variability in the ground gas that can occur the actual gas concentration in the wells could vary between each monitoring event, although the significance of this will vary depending on site specific factors such as the ground conditions and the period between monitoring events.

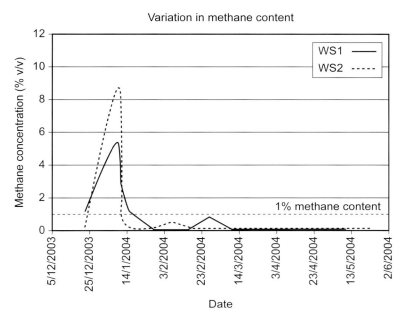

Figure 6.2 *Early peaks in gas concentration*

It is also useful to plot graphs of methane against carbon dioxide concentration. Where there is significant active gas generation in the ground around the borehole the ratio of $CH_4:CO_2$ should be about 60%:40%. Graphs of methane and borehole flow against atmospheric pressure and water levels help to identify if there is any correlation between the gas regime and external factors.

It is not very common to observe a relationship between atmospheric pressure and gas concentrations in wells that are installed in a low generation material.

A comparison of the results of the gas monitoring with ground conditions can help to identify the main source of gas on a site. For example, it is often useful to group together the results of gas concentration or borehole flow rate for wells with response zones in a particular soil type.

6.3.3 Plotting contours

Contours of gas concentrations can be used to identify any migration or variation in the gas regime (see Figure 6.3). However, contours of gas readings should only use readings from same time and day. Ground gas is mobile and the regime will change with time in response to external factors. Therefore gas data used in contouring should all be taken from the same date otherwise it could produce misleading results (for example do not use the maximum gas concentration from each borehole, regardless of when the maximum occurred).

To be able to plot meaningful contours sufficient data will be required to make this a valid exercise. Care should be taken when using interpolated data

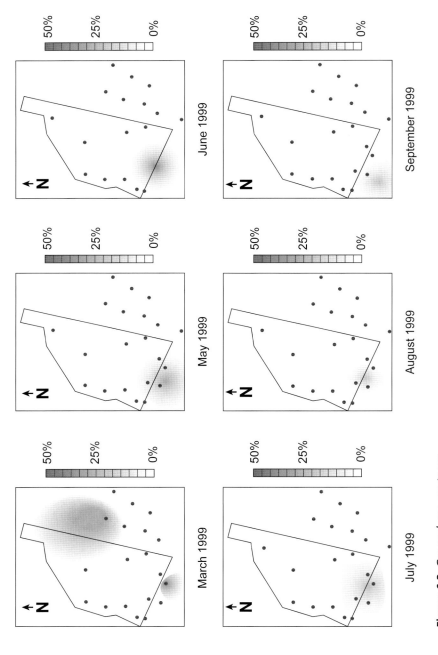

Figure 6.3 *Ground gas contours*

and modelling, especially on boundaries, as it can result in misleading diagrams that do not replicate what is happening in practice. Contours should be given a 'reality check' to make sure that they make sense when other data is considered.

6.3.4 Statistical analysis

Where there is a suitable comprehensive data set it may be useful to calculate the mean and standard deviation of the readings. This helps to determine the significance of the maximum results obtained on a site and the presence of outliers in the data set. If the group of higher results are clustered then consideration can be given to zoning this area of the site and assigning a high characteristic situation and hence adopting greater gas protection measures. However, this would require the higher results to be correlated with a difference in ground conditions.

6.3.5 Temperature

The temperature inside monitoring wells is not routinely measured as part of most ground gas investigations. However, it is a useful parameter to record. In sources where there is significant gas generation occurring (e.g. deeper domestic refuse landfill sites) temperatures can reach over 60°C, with 40–45°C common in the first five years after landfilling. These values may lead to increased headspace temperatures in the monitoring wells. Shallow landfills may be more sensitive to climatic conditions than deeper ones and landfill gas production may tend to drop when the temperature falls below 10–15°C. This may result in a seasonal pattern of waste decomposition and gas production.

6.3.6 As-built conditions

It is important during any ground gas assessment to consider whether the data is representative of as-built conditions. For example, is there a surface clay layer overlying a gas source that will be penetrated by stone columns or piles that allow gas to migrate to surface? This is common on sites with alluvium where a clay layer overlies peat deposits at depth. In this case any assessment needs to consider the gas regime in the deeper source and how gas will migrate up the stone columns.

6.4 Analysis of vapour data

Much of the preceding advice about analysing gas data can also be applied to vapour data. Vapour data can also play an important role in validating the results of risk assessments that are based on soil and groundwater concentrations of volatile contaminants.

6.5 Tier 1 risk assessment: Is there a linkage?

The first part of the gas risk assessment is to identify any source pathway and receptor linkages from the conceptual model. If there is no linkage then there is no risk. In Figure 6.4 there is no linkage for bulk flow of methane and carbon dioxide between the gassing ground on the left-hand side (a former land raise) and the housing on the right-hand side. This is because the clay soils below the land raise prevent migration through the ground and the presence of the ditch between the two is, in effect, a cut-off barrier. However, there is a linkage for the transmission of vapours emitted from the surface of the ground to migrate to the houses via the atmosphere.

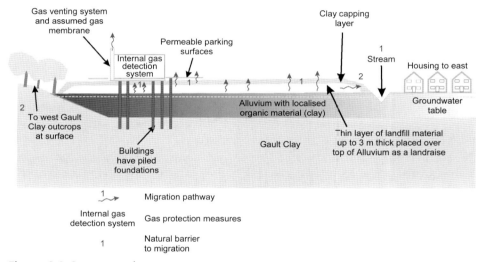

Figure 6.4 *Source–pathway–receptor*

Once the source–pathway–receptor scenarios have been identified the risks associated with each one can be assessed qualitatively (i.e. descriptively). Guidance on assigning the likelihood and consequences to the scenarios is given in Tables 6.1–6.3.

Table 6.1 *Categories of likelihood (Environment Agency, 2004)*

Description	Probability
Probable	Occurs at least once per year
Fairly probable	Occurs between once per 10 years and once per year
Somewhat likely	Occurs between once per 100 years and once per 10 years
Unlikely	Occurs between once per 10,000 years and once per 100 years
Very unlikely	Occurs between once per million years and once per 10,000 years
Extremely unlikely	Occurs less than once in a million years

Table 6.2 *Categories of consequence*

Description	Example scenario
Severe	Short-term or acute risk of serious harm to human health or death. Explosion in a building or asphyxiation of occupants
Substantial	Chronic damage to human health due to long term exposure to vapours or gases. Explosion in open space that is inaccessible to public
Marginal	Cause odours or financial blight to a development
Negligible	No effects to human health. Easily repairable damage to buildings or other structures Presence of gas at such concentrations that PPE is required for workers on site

Table 6.3 *Comparison of consequence against likelihood*

		Consequence			
		Severe	Substantial	Marginal	Negligible
Category of probability	Probable	Very high risk	High risk	Moderate risk	Moderate/low risk
	Fairly probable	High risk	Moderate risk	Moderate/low risk	Low risk
	Somewhat likely	Moderate risk	Moderate/low risk	Low risk	Very low risk
	Unlikely	Moderate/low risk	Low risk	Very low risk	Very low risk
	Very unlikely	Low risk	Very low risk	Very low risk	Negligible risk
	Extremely unlikely	Very low risk	Very low risk	Negligible risk	Negligible risk

Notes

[1] **Very high risk**: There is a high probability that severe harm could arise to a designated receptor from an identified hazard, or there is evidence that severe harm to a designated receptor is currently happening. This risk if realised is likely to result in substantial liability

[2] **High risk**: Harm is likely to arise to a designated receptor from an identified hazard. This risk if realised is likely to result in substantial liability

[3] **Moderate risk**: It is possible that harm could arise to a designated receptor from an identified hazard. However it is either relatively unlikely that any such harm would occur or if any harm were to occur it is likely that the harm would be mild

[4] **Low risk**: It is possible that harm could arise to a designated receptor from an identified hazard but it is likely that if realised it would at worst be mild

[5] **Very low risk**: There is a low possibility that harm could arise to a receptor and in the event of harm being realised it is not likely to be severe

[6] **Negligible risk**: It is beyond the realms of credibility that harm could occur to a receptor and if it is realised the effects will be marginal

As an example of the application of Table 6.2 consider a house that has poorly ventilated or confined spaces. The consequence of migration of gas into the house may be considered to be severe. However, for a warehouse that is a large well-ventilated open space the consequence of gas migration may be considered to be marginal. An example of the application of these descriptions to various pollutant linkages is provided in Box 6.1.

Box 6.1 Examples of source–pathway–receptor assessments

Pollutant link-age	Estimated probability	Estimated consequence	Estimated risk (combination of conse-quence and probability)	Mitigation
Migration of gas into hous-ing over landfill site	Fairly probable	Severe	High risk	Develop site with flats only with extensive gas protec-tion measures and no private gardens
Widespread migration to existing hous-ing outside landfill site	Unlikely	Severe	Moderate/low risk	Provide venting trench
Widespread migration to existing park-land outside landfill site	Unlikely	Marginal	Very low risk	None
Migration into supermarket	Unlikely	Substantial	Low risk	Provide gas protection to supermarket building
Gas migration from made ground and hydrocarbon contamination into housing	Very unlikely	Severe	Low risk	Provide gas protection to new houses outside landfill sites
Gas migration from alluvium into housing	Extremely unlikely	Severe	Very low risk	None

Mitigation is normally required where the estimated risk is low or greater in order to reduce levels to low or very low.

At this stage, if historical gas monitoring results are available that are representative of current ground conditions it may also be useful to compare them against generic screening values. Historical data often lacks borehole flow rates but the gas concentrations can be used in a simple screening assessment to determine if the gas regime below a site is below the level of concern such that there is no requirement for gas protection measures.

These criteria can be based on the following:

- 2.5% methane and 5% carbon dioxide for low sensitivity (e.g. industrial/commercial buildings with open spaces and substantial engineered floor slabs). This is based on guidance by Pecksen (1986)
- 1% methane and 5% carbon dioxide for housing or other high sensitivity end use

It should be noted that if gas concentrations exceed these levels it does not mean that the risk is unacceptable. It simply means that further assessment will be required to determine if the proposed development is suitable and what level of gas protection will be required. They do, however, give a quick and easy means of screening out very low risk sites (see Box 6.2).

Box 6.2 Screening with historical gas data

A site was underlain by 2 m of made ground that was inert soil with very occasional pieces of wood in it. The site was to be developed with a steel framed warehouse building with piled foundations and a reinforced concrete cast in situ suspended floor slab. Historical gas monitoring data was available but did not include flow rates.

The gas monitoring data indicated that generally methane was not detected, except for occasional readings up to 1.9% v/v in one borehole out of five. Carbon dioxide concentrations up to 3.1% were present sporadically in certain wells.

It was concluded that the results were representative of the site conditions and were below the screening levels of 2.5% methane and 5% carbon dioxide. Therefore gas protection measures were not required.

6.6 Tier 2 risk assessment: qualitative assessment for methane and carbon dioxide

6.6.1 Gas screening value

CIRIA Report C659/665 has two methods of characterising the level of risk associated with a site, one that has been specifically developed for the NHBC that applies for low rise housing and the other for any other type of development. Both methods use the gas screening value (GSV) to characterise a site. The GSV is given by:

$$GSV = \text{borehole flow rate (l/h)} \times \text{gas concentration (\%)}$$

It should be noted that the gas concentration is the mathematical form (e.g. 50% is 0.5 in the equation).

The GSV is based on either methane or carbon dioxide concentrations. Trace gases and vapours are often at much lower concentrations. So the classification and design of protection measures based on methane and carbon dioxide flows is usually more than adequate to provide protection against the acute effects from other gases. However, a human health risk assessment is also likely to be required to assess the chronic effects of trace gases and vapours. The human health risk assessment may indicate that remediation of the trace gases or vapours is required in addition to any building protection measures. The human health risk assessment may also need to consider the emissions from any underfloor venting system.

Given that the most onerous concentration and borehole flow values may never be identified directly from a set of borehole gas monitoring readings (see Chapter 5) it is far better to derive a set of design parameters using a appropriate factor of safety. Thus the GSV can be estimated using two approaches that are discussed below.

The first approach to calculating the GSV is by adopting a worst credible approach to assess the risk from soil gas. This is achieved by calculating the GSV for each set of gas monitoring readings and then taking the maximum value of GSV obtained. A factor of safety may be applied to this, depending on how comprehensive the data is. This design value is then used to classify the site and define the scope of the gas protection measures. The design value is also used in the engineering design of the gas control measures, particularly the design of ventilation systems.

An alternative, very conservative, approach adopted by some landfill gas practitioners is to determine the GSV using a worst possible approach combined with a factor of safety of one. In this approach the worst combination of parameters from the data set, irrespective of whether they can physically occur simultaneously, is used to calculate the GSV (i.e. the highest flow rate and highest gas concentration measured anywhere on a site). In this way there is an inherent high margin of safety built into the data and determination of the site characteristic situation. This approach is usually adopted where there is a limited data set. (See Box 6.3)

Box 6.3 Calculation of GSV

Worst credible approach
The GSV calculated from all the readings on a site is shown in Figure 6.5.

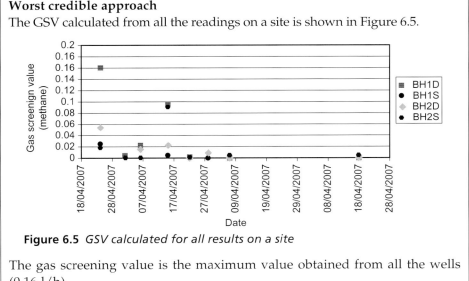

Figure 6.5 *GSV calculated for all results on a site*

The gas screening value is the maximum value obtained from all the wells (0.16 l/h).

Worst possible approach
Gas monitoring on a site has given the following readings:

gas concentration = 1.2–45% (worst case = 45%)
borehole flow rate reading from gas meter = 0.1–3 l/h (worst case = 3.1l/h)

GSV = 3 l/h × 45/100 = 1.35 l/h

6.7 Screening process

The screening process uses the GSV to determine the characteristic situation of a site (Wilson and Card) or the traffic light classification (NHBC). The choice of appropriate method is shown in Figure 6.6. This approach is consistent with the guidance in CIRIA Report C659/665 which indicates that the NHBC approach is only for developments with houses that have gardens, a suspended floor slab and a ventilated underfloor void (Situation B in CIRIA C659/665). For all other development types Situation A should be used.

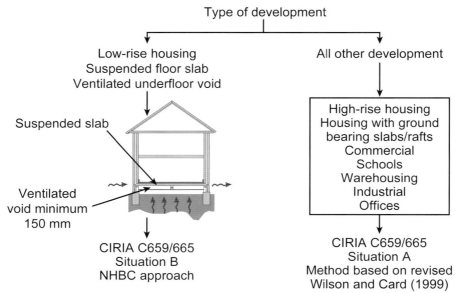

Figure 6.6 *Flow chart for Tier 2 assessment*

It should be noted that it is often useful to use both the NHBC method and the Wilson and Card method for low-rise housing with a void as a sensitivity check on the results.

6.7.1 NHBC approach (Situation B in CIRIA C659/665)

For low-rise housing with gardens, where a ventilated underfloor void is provided, the GSV is used in Table 6.4 (see colour section) to determine the level of risk (traffic light classification: green to red).

Once the site has been classified the scope of protection measures can be determined during the design process (described in Chapter 8). It should be noted that the classification system assumes there will be an underfloor ventilated void as the very minimum of protection (even for the green classification).

6.7.2 Wilson and Card approach (Situation A in CIRIA C659/665)

For other types of development the site is characterised using the GSV to give a characteristic situation using Table 6.5 (see colour section), which is a modified version of the table from Wilson and Card (1999). Other types of development includes low-rise housing with raft foundations, suspended cast in situ slabs or ground-bearing slabs and also high-rise housing, commercial and industrial developments, schools etc.

Again, once the level of risk is identified the scope of protection measures can be defined following the procedures in Chapter 8.

6.7.3 Differences between Situation A and Situation B methods

The Situation A method is based on the approach adopted by Wilson and Card which was based on a study of the range of gas protection measures installed in various sites and the performance of underfloor ventilation systems. It is therefore an empirical approach based on previous practice and observed performance. The main principle is that, as the level of risk increases the number of individual components in the protection system increases. Each component is capable of protecting a building on its own should the other components be damaged or stop working for whatever reason. It assumes that each component will be designed and specified to acceptable standards (e.g. design of venting to achieve very good performance in accordance with the Partners in Technology Report (see Chapter 8).

The Situation B approach (Boyle and Witherington, 2007) is based on an analysis of gas venting below a typical house suspended floor slab and the build up of gas within the void and a typical small room in the building above. It is, however, dependent on factors such as: the height of the void, air changes within the void etc. It is important when assessing a site with this method that the development fits the assumptions made in the analysis, including an allowance for blocking of vents etc. by householders.

The Situation B approach also has more limitations placed on each classification in terms of gas concentrations as well as GSVs.

The two methods have been derived from different principles and the boundaries between the classifications will differ. However, it has been found in practice that for the vast majority of sites where housing is proposed, there is little difference in the recommendations obtained from each method. Most sites are within the limits of Characteristic Situation 2 or Amber 1, which requires a gas membrane and underfloor venting in both cases. The Situation A approach is slightly more stringent on gas membrane requirements and the Situation B is more stringent in terms of limiting gas concentrations.

The Situation B method limits housing to sites that are up to Amber 2 unless the risk assessment can demonstrate that housing would be acceptable. The Situation A method does not do this, but it does require a comprehensive qualitative risk assessment to justify the gas protection measures for any characteristic situation above 3. The two methods are therefore similar in this approach.

The Situation B approach has different limits for carbon dioxide than for methane whereas the Situation A approach has the same limits regardless of the gas. This is because the risk posed by carbon dioxide is not considered to be any less than that posed by methane. Since acute adverse health effects occur at 1.5% carbon dioxide, it could be argued that this is more critical. Box 6.4 (see colour section) compares the two methods of site classification.

6.8 Tier 3 risk assessment: quantitative assessment for methane and carbon dioxide

A quantitative risk assessment is rarely required except on the most difficult or sensitive of sites. Sites that are being assessed under Part 2A of the Environmental Protection Act 1990 are likely to require a quantitative assessment to provide a robust indication that there is a pollutant linkage.

The most commonly used method of quantitatively assessing risk on gassing sites is that described in CIRIA Report 152. This uses a fault tree analysis to provide a numerical estimate of the risk (i.e. a probability that an adverse effect will occur in any year or other specified period). The method described below varies from the method described in CIRIA 152, based on published papers (Hartless, 2004; Sladen and Dorrell, 2001) and the present authors' experience.

The fault tree is used to estimate the risk of gas migrating into a building and accumulating to an explosive concentration. This is the probability of a hazard occurring. Once the probability of the hazard occurring is known then the frequency of exposure to the hazard (usually the frequency of ignition in a methane assessment) must be assessed. The overall risk is then given by:

risk = frequency of exposure to hazard × probability of hazard occurring

A fault tree is developed by working backwards from the hazard or undesirable event (the top event) and listing in reverse order the things that must happen for the hazard to be realised (contributory events). This continues until a basic event is arrived at where no further events can be developed. Probabilities are estimated for the subevents and then the overall probability of the top event occurring can be estimated.

The vents in a fault tree are connected by logic gates and most fault tree analyses for gas protection systems can be carried out using only the following symbols:

(1) **TOP event**: forseeable, undesirable event
(2) **Intermediate or contributory event**: events that must happen for top event to occur
(3) **Basic event**: Initiating fault/ failures that are not developed further. The basic event marks the limit of resolution of the analysis
(4) **OR gate**: produces output if any input exists
(5) **AND gate**: produces output if all inputs co-exist

The basic concept of developing a fault tree is shown in Figure 6.7. An example showing the development of a simple fault tree for a methane explosion in a building is provided in Figure 6.8. The complete fault tree is shown in Figure 6.9.

Once the fault tree has been completed the probability of each event occurring can be estimated and thus the overall probability for the top event can be calculated.

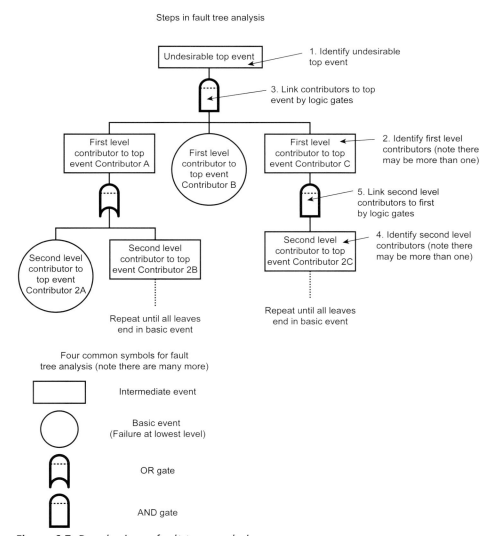

Figure 6.7 *Developing a fault tree analysis*

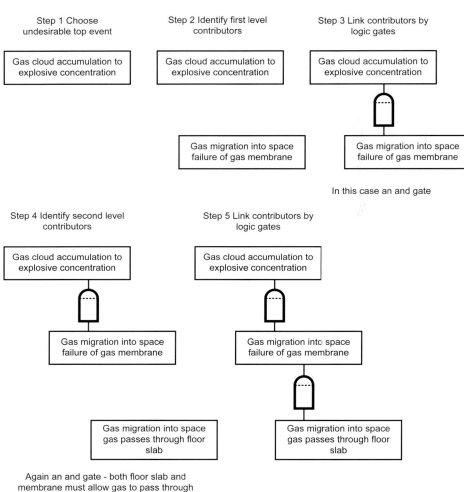

Figure 6.8 *Example development of fault tree*

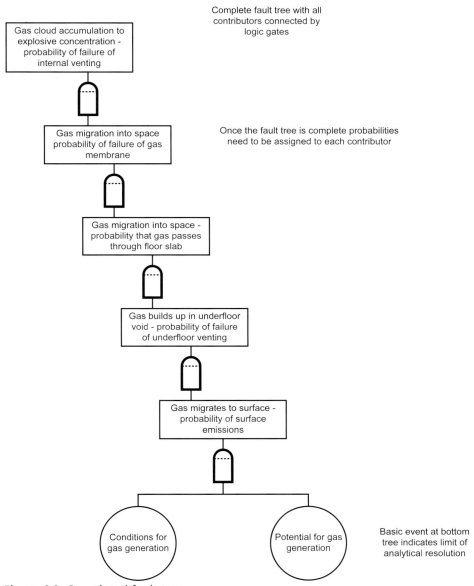

Figure 6.9 *Completed fault tree*

Useful methods of defining probabilities for some of the events have been given by Hartless (2004). These are summarised below and correct some of the errors that have been noted in the method described in CIRIA Report 152 (CIRIA, 1995).

6.8.1 Entry of gas into building

Calculation of gas entry rates into buildings is an area where there is the greatest degree of uncertainty when undertaking risk calculations. When gas protection is provided the gas entry is closely linked with the performance of both the sub-floor ventilation measures and the gas resistant membrane.

The method described in CIRIA Report 152 assumed that the actual entry rate of methane into the cupboard was the diffusion rate of methane from the adjacent landfill. However, this possibly does not adequately represent the rate at which methane will enter a cupboard. A more realistic indicator of methane surface emissions into the cupboard can be gained using borehole flow rates or calculated surface emission rates. One method of doing this, if there is a sufficient data set, is by undertaking statistical analysis of the gas monitoring results (see Box 6.5).

Box 6.5 Statistical analysis of gas monitoring results

The results can be plotted as a frequency distribution to determine the modal value. Normally the frequency distribution is based on ranges of gas values as shown in Figure 6.10.

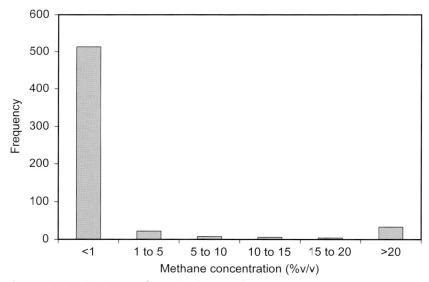

Figure 6.10 *Histogram of monitoring results*

The frequency distribution can also be used to estimate the type of probability distribution that applies to a data set. For a continuous function, the probability density function (pdf) is the probability that the variate has the value x, thus the probability of a parameter exceeding a particular value can be estimated. For example the probability that gas emissions overwhelm an under floor venting system can be estimated.

From the above graph:

$$f(x) = \frac{1}{\mu} e^{\frac{-x}{\mu}}$$

and

$$P(x < x_0) = 1 - e^{\frac{-x_0}{\mu}}$$

Where μ = mean of data. So the probability that the methane concentration is less than 1% from the above graph (mean of data is 2%) is given by:

$P(<1\%) = 1 - e^{-1/2} = 0.39$

A similar process can be undertaken for differently shaped distributions.

In CIRIA Report 152 the construction considered was an in situ concrete floor slab incorporating a HDPE membrane. The report went on to postulate that on the basis of plan areas there was a probability of 0.001 that there would be a crack in the floor slab beneath a cupboard. This was just a hypothesis. Any assessment of the risk of membrane failure will be based on judgment as there are some factors that are difficult to apply numbers to (e.g. the risk of a worker drilling through a membrane). However, the scoring method described in Box 6.6 can help as an aid to judgment.

Box 6.6 Estimating the probability of membrane damage

The probability of damage occurring to a membrane can be estimated based on data produced by the Environment Agency for landfill liner systems (Environment Agency (2004a), The likely medium- to long term generation of defects in geomembrane liners. R&D Technical Report P1-500/1/ TR). Using this guidance the base pdf for the number of defects per hectare of a membrane liner during installation with good quality control is summarised below.

Estimated hole frequencies per hectare of liner (Environment Agency, 2004a)

Category	Pinholes (0.1–5 mm²)			Holes (5–100 mm²)			Tears (100–10,000 mm²)			Total		
	Minimum	Most likely	Maximum	Minimum	Most likely	Maximum	Minimum	Most likely	Maximum	Minimum	Most likely	Maximum
Good case	0	10	15	0	5	10	0	2	5	0	17	30

The base probability is adjusted up or down depending upon the site and material specific factors that affect the probability of membrane damage on any development.

Factor	Adjustment to probability
Design life: as design life increases the risk of damage occurring increases	Increase probability of failure by 10% for every 20 years of design life over 50 years
Construction Quality Assurance (CQA): if a membrane is installed under a CQA system and joints are tested the risk of defects is reduced, conversely without a CQA system the risk of installation defects increases	If membrane installed without CQA and seam testing increase probability of defects by 300%

If membrane is installed without seam testing but with CQA increase probability of defects by 150% |
Specialist installers: if specialist installers are used the risk of construction defects is reduced and conversely if the membrane is installed by ground workers, bricklayers or other similar staff the risk of defects is increased	If membrane is not installed by specialists increase probability of failure by 100%
Independent inspection: if the membrane is inspected by an independent consultant immediately before screeding or concreting the risk of construction defects is reduced	If there is no independent inspection immediately before covering the membrane increase the probability of defects by 200%
Area of membrane: smaller areas have a proportionately greater number of seams. Landfill liner data is based on large areas of installation and most developments will require smaller areas	Increased probability of failure by 200% for domestic installations and 100% for large commercial or industrial installations
Complex details: number of service penetrations or complex structural slab and foundation details	If there are complex foundation details that membrane must be sealed to or a large number of service penetrations then increase probability of failure by 300%
Type of membrane: thin damp proof membranes (DPMs) are usually made of low density polyethylene (LDPE) and will have a greater risk of defects than specific gas-resistant membranes manufactured from more robust materials (polypropylene, linear low density polyethylene and high density polyethylene). As the strength and puncture resistance decrease the probability of defects increases	For thin DPM material increase probability of failure defects

2000 g–200%
1200 g–400%
1000 g–600%

Note these values do not apply to specific gas resistant membranes made from more robust materials than DPMs |
| Protective layer: if a protective layer such as a geotextile fleece or sand layer is not provided the risk of defects increases | If a protective layer is not provided increase probability of failure by 200% |

Factor	Adjustment to probability
Settlement: the likely differential settlement between building components will affect the risk of membrane failure. The greater the level of likely movement the greater the risk	Site-specific assessment based on likely level of movement. High elongation membranes reduce the risk of total failure
Position in construction: membranes placed below a concrete slab face less risk of accidental damage by the occupiers than those above the slab and below a screed	If membrane is above structural floor slab increase probability of damage by 300%
Development: the risk of accidental membrane damage is greater in houses than in large commercial buildings	For houses increase risk of accidental damage by 300%

Where there are significant gaps in the floor slab (e.g. service entry points) the probability of failure would be 1.0. This is particularly relevant in block and beam slabs where the service entry points are often poorly sealed. It is less of an issue in large building with in situ reinforced concrete slabs with cast in service entries.

6.8.2 Failure of subfloor ventilation

This topic is not covered in CIRIA Report 152. There are different methods of venting underfloor voids and three common scenarios are:

- Calculate passive ventilation rates (driven by wind effects) to see if it is adequate to disperse the incoming gas
- Consider still wind situations where there is no ventilation
- Calculate the probability of failure of active pumping systems using the mean time between failure (MTBF) data for fans

The calculations for the first two scenarios are discussed in more detail in Chapter 7. A method of calculating the probability of pump failure for an active system is shown in Box 6.7.

Box 6.7 Example calculation: probability of fan failure for active system

Assume mean time between failure (MTBF) of fans = 25,000 run time hours.
MTBF = $1/\lambda$ so $\lambda = 25000/(24 \times 365) = 1/2.85$ yrs
Where λ = failure rate. Assume that the Poisson distribution to failure and constant rate of failure (optimistic as rate of failure increases with run time as fans wear out) is:
probability of one failure in a year $P(X = 1)$
$= [(\lambda t)^x \, e^{(-\lambda t)}]/x! = 0.16$
Where t = time period (1 year in this case).

Hartless (2004) has demonstrated that in most cases for methane to reach explosive levels within a passively ventilated void there would need to be a significant failure of the passive ventilation (for example if all the vents become blocked) or the entry rate of gas would have to increase by orders of magnitude. For simplicity he assumed that the probability of such an occurrence was 0.01. One approach that is less subjective is to estimate the failure rate using the probability density method described previously (Box 6.4, see colour section). An example is shown in Box 6.8. However, it is difficult to allow for human factors when estimating the probability of failure for elements of a protective system and an element of professional judgment will always be required. Studying examples and guidance from other industries can also assist in this respect (e.g. chemical and oil industries).

Box 6.8 Estimating probability of failure of underfloor venting

This example describes an approach to estimating probability. The assessor should ensure that it is appropriate to the specific conditions on any particular site.

The underfloor venting has been determined by calculation (see Chapter 7) to be effective up to a borehole flow rate of 1.6 l/h. We can estimate the probability of this value being exceeded (assuming the gas concentration remains constant). The example also assumes that all the data is representative of the gas source on the site (any data from boreholes that are not representative should not be included in the analysis).

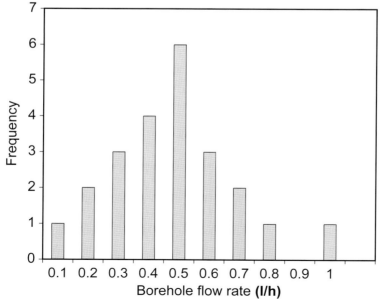

Figure 6.11 *Borehole flow rate and pdf for normal distribution*

The pdf for borehole flow rates measured in boreholes in the site is given in this case by a normal distribution shown in Figure 6.11 (note that the pdf chosen must be appropriate to the data being analysed).

For a normal distribution the pdf is given by:

$$P_n = \frac{1}{\sigma\sqrt{2\pi}} e^{[-(x-\mu)^2/2\sigma^2)]}$$

Where:

μ = mean of data (in this case = 0.5 l/h)
σ = standard deviation of data (in this case = 0.22 l/h)
P_n = probability that borehole flow rate = n

So the probability that the borehole flow rate is greater than 1.6 l/h from the above graph (mean of data is 0.5 l/h, standard deviation = 0.22 l/h) is given by:

1.6 l/h = $\mu + 5\sigma$ and from the erf function the probability that a result lies within $\mu \pm 5\sigma$ is 0.9999994. (From Figure 6.11 it can be seen that the probability that a value lies within a certain range increases the further to the right the value is on the graph. See any textbook on statistics for a full explanation of this derivation.)

The probability that the flow rate is greater than 1.6 l/h is (1 – 0.999,9994) / 2 = 3 × 10⁻⁷.

6.8.3 Justification of parameters and sensitivity analysis

In order to help make the quantitative risk assessment process transparent and easy to understand the choice of parameters should be clearly explained and recorded. A useful method of justifying the baseline input parameters and also defining the likely sensitivity of each one is summarised in Table 6.6.

Table 6.6 *Example of baseline and sensitivity parameters*

Parameter	Baseline value	Range for sensitivity analysis	Justification
Methane concentration	22% v/v	6.7–8.5% v/v	To cause migration concentration must be consistent over period of migration. Gas levels are variable and using maximum values is unrealistic as they do not occur all over site or all the time. So use mean value and check with maximum values in sensitivity analysis. Note that this approach requires a high density data set and possibly continuous monitoring which may help to identify a baseline value

Parameter	Baseline value	Range for sensitivity analysis	Justification
Carbon dioxide concentration and oxygen depletion	N/a	N/a	The critical level of methane is 1% v/v. The critical level for carbon dioxide is 1.5% v/v and for oxygen depletion any gas must remove oxygen so that levels fall to 17% v/v or below before effects take place (CIRIA Report 149). This requires more than 1% v/v of gas. So worse case is 1% v/v design criteria.
Borehole flow rate	13.6 l/h	N/a	Borehole flow rate is maximum recorded in the site and therefore no sensitivity analysis has been used for this parameter
Gas pressure	20 Pa	N/a	Value is very worse case based on experience of monitoring pressures in similar types of material
Atmospheric pressure drop	2900 Pa	N/a	Worse case value that occurred (over 7 h) at Loscoe explosion
Thickness of migration pathway	4 m	N/a	Worse case: maximum thickness of made ground below site (this also includes the original made ground below the made ground placed as part of the development, so is the very worse case)
Permeability of soil to water (hydraulic conductivity)	1×10^{-5} m/s	1×10^{-3} m/s to 1×10^{-7} m/s	Soils described as clayey silty sands. The permeability will be governed by fine content. G N Smith gives ranges of permeability and baseline value is worse case for fine sand or silt at $1 \times 10{-5}$ m/s but could be as low as 1×10^{-7} m/s. This will be converted to intrinsic permeability for analysis of gas flow
Diffusion coefficient for soil	1×10^{-6} m²/s	N/a	This value is from CIRIA Report 152 for a soil with a voids ratio of 0.25. The soils in this site are unlikely to have a voids ratio greater than 0.3 and the table in CIRIA 152 indicates this will not significantly affect the diffusion coefficient
Viscosity of ground gas	1.03×10^{-5} Ns/m²	N/a	From CIRIA Report 152 value for methane

Parameter	Baseline value	Range for sensitivity analysis	Justification
Density of ground gas	0.717 kg/m³	N/a	From CIRIA Report 152 value for methane
Ventilation of confined space in build- ing	2 air changes per day	N/a	Worse case value based on CIRIA 152 and Sladen et al. (2001)
Defects in gas Membranes	17 defects per Ha	30 defects per Ha	Membranes were installed by special- ist subcontractor, under QA system and independently verified
Size of house	6 m × 8 m	N/a	Typical size of individual unit from layout plans. Longer terraces will have negligible effect on calculations
Size of cup- board	1 m × 1m × 2 m	N/a	Cupboard with light switch or other electrical source of ignition. From floor plans of development

Sensitivity analysis is important to identify if any variations in a particular parameter have a significant effect on the results. Figure 6.12 shows the variations in gas concentrations being assessed to see their effect on the estimated risk.

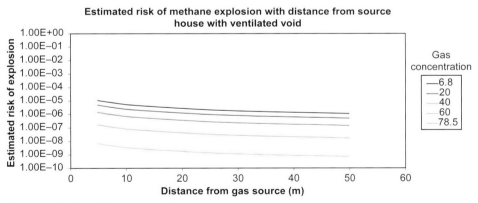

Figure 6.12 *Sensitivity analysis*

In this case the results of the risk assessment showed that for the houses located more than 50 m from the edge of the gas source there was negligible risk (less than 1×10^{-6}) from ground gas migration, even when using the highest gas concentration ever recorded on the site. Therefore it was con- sidered that gas protection (over and above a ventilated underfloor void) was unlikely to be necessary for the houses that were more than 50 m from the source. At a distances between 5 m and 50 m from the source the risk could exceed 1×10^{-6} when considering the higher values of gas concentration that were measured in the source. In this case, it was concluded that the provision of a gas membrane would reduce the risk to an acceptable level. It should be noted that these judgements were based on a great deal of analysis of the data for the site, including likely generation rates. The sensitivity analysis was only one part of a detailed assessment that allowed these conclusion to be drawn.

6.8.4 Limitations and considerations of fault tree analysis

Fault tree analysis is highly effective in determining how combinations of events and failures can cause specific system failures, however, the technique has three main limitations:

- Fault tree analysis examines only one specific event of interest and so has a narrow focus. To analyse other events, other fault trees must be developed
- Fault tree analysis is as much an art as a science and a significant degree of judgement is required. The level of detail, types of events included and organisation of the tree vary significantly between analysts. However, given the same scope of analysis and limiting assumptions, different analysts should produce comparable, if not identical, results
- Using fault tree analysis results to make statistical predictions about future system performance is complex and requires a great deal of experience. It is a complex and specialist field and anyone undertaking an analysis should be fully conversant with the application of the technique and be aware of how it is applied in other fields

The focus of many risk assessments often becomes concentrated on equipment and systems and human and organisational issues are not adequately addressed. For example, in gas risk assessments how can the risk of accidental membrane damage or blocking of vents by residents in a housing development be accurately predicted? There will always be an element of subjectivity in assigning probabilities to these aspects. It is therefore useful to discuss and agree values for these parameters with the regulators, prior to undertaking a detailed analysis and to achieve a consensus view on the results.

An important concept of fault tree analysis is that all the events and subevents are independent of each other. So for example in the CIRIA 152 fault tree the cause of ventilation cannot be dependent on a factor that also causes ignition (i.e. entry into the cupboard).

Fault tree analysis can be useful when undertaking Part 2A risk assessment for ground gas. Gas migration can be modelled using the equations and approaches discussed in Chapter 4. In particular the effect of providing gas barriers around sites can be assessed.

6.8.5 Common mistakes in fault tree risk assessments

Common mistakes or omissions when undertaking fault tree analysis are:

- There should be a comprehensive justification of parameters and sensitivity analysis
- Ignition should be outside the fault tree
- The risk assessment should be site specific and be related to the actual development
- The risk assessment should be undertaken without gas protection measures and then developed to identify the most appropriate combination of protection measures. A common fault is to use the risk assessment to define the need for protection and then just state that protection will be provided
- Undue reliance on the numbers: remember they are only an aid to judgement

6.9 Assessments for house extensions

Assessing the risk posed by gas to extensions to existing buildings is difficult because, historically, many houses have been built without any gas protection on sites with low levels of gas. In those cases it is difficult to justify adding gas protection to only the new part of a house unless there is a clearly known risk. It is also unreasonable in many cases to ask householders to carry out desk studies or monitoring for ground gas. A suggested approach is given in Box 6.9.

Box 6.9 Suggested guidance on house extensions

Landfill and ground gas can be a potential hazard to health. If there are plans to make structural alterations to a property that is within 250 m of a known landfill site or within an area of suspected ground gas it is likely that the local council will require gas protection measures to be installed.

Protection to extensions is generally required if:

- Existing building has protection measures
- There are known problems with gas entering unprotected buildings in the area
- There is a high-risk landfill site nearby and gas migration is known to be occurring from it close to the property being considered.

In other situations where there is the potential for gas but the risk is low and the existing property does not have gas protection measures then protection to the extension may not be required.

If the property already has gas protection measures or the local council has put a condition requiring gas protection measures on the planning approval for the extension it is important that such measures are installed. Such a condition will only be placed on a planning approval if the council believes that the property and its occupants may be at risk without these protection measures.

If a building contains a certain specification of ground gas protection measures then an extension should not be constructed without protection or with measures of a lower specification. Doing so may compromise the gas protection measures to the whole property. Therefore it is important that every property extension this is constructed should contain gas protection measures to at least the same specification of that in the existing building.

It should also be ensured that the construction of extensions does not in any way compromise the effectiveness of existing gas protection measures (e.g. by blocking vents to the underfloor void or damaging existing membranes).

In cases where protection is required for an extension it should be possible to design it without gas monitoring data and site investigation data. In most situations it will comprise a membrane and possibly underfloor venting.

6.10 Putting ground gas risk in context

It is interesting to note that historically, housing in many areas of the UK has been constructed on peat without gas protection measures. The present authors are not aware of any incident where ground gas ingress from deposits of peat has caused an explosion. Even though methane concentrations in peat may be high the flow rates are usually low since bulk gas production has ceased. Methane will migrate slowly towards the ground surface and is largely oxidised to carbon dioxide near the surface in aerobic soil conditions. However, foundation construction can affect this equilibrium (e.g. by piling or ground improvement) and release large volumes of trapped gas, sometimes at high concentrations. They can also provide a preferential pathway for gas migration in the long term. As such due care and attention should be paid during construction with respect to confined spaces etc.

Sewers and oil separators are often put forward as a source of ground gas and vapours. However, modern prefabricated oil separators are well vented and sealed and the risk of migration from these through the ground is negligible.

Although sewers can have methane in them, it is virtually impossible for it to travel any distance through the ground outside the sewer (except in the backfill) because sufficient pressure cannot be generated to cause migration (it would cause other problems within the system before migration in the ground occurred).

It is also useful to compare the risk posed by the presence of the gas in the ground to other sources of gas within buildings. There are other sources that will emit gases directly into the building:

- Humans (methane and carbon dioxide)
- Cars in car parks (carbon monoxide)
- Gas-fired heating and cooking appliances

Calculations on one site indicated that the rate of emission of methane from human sources (flatulence) directly into the occupied spaces could quite feasibly be at a rate of up to 4.5 l/day. This was similar to the rate of gas emission that was likely from the low risk sources below the site. As far as the present authors are aware there is no recorded case of methane explosions occurring due to human sources. Similarly carbon dioxide from respiration could be emitted directly into the occupied spaces at a rate of 315 l/day.

Cars emit carbon monoxide from their exhausts. This is greatest when the car has started and the catalytic converter has not warmed up (typical car park conditions). Carbon monoxide is a highly toxic gas that is also explosive when present between 12.5–74% by volume in air (thus it is explosive in a wider range of concentrations than methane, although the lower explosive level is higher). It can be emitted from car exhausts at concentrations in the range 0.5–3.5% by volume, depending on the year of manufacture and condition of the vehicle. Badly maintained cars will exceed these limits. Car park ventilation design is based on allowing an inflow of carbon monoxide of up to 2.52 m^3/h per vehicle with an engine running. This will give far greater emission rates directly into the parking area than are likely from of migration of gas from the ground on many sites where only low generation sources are present.

Similarly gas fired appliances generate carbon dioxide within the occupied space at a greater rate than emissions from the ground on many low risk sites.

The ventilation in car parks and buildings is more than adequate to dilute these other sources of gas. Since these other gases are emitted directly into the building the risks posed by them are far greater than the risk posed by the presence of gas in the ground on many sites where low surface emission rates are predicted.

6.11 Nuisance from ground gas: odours

In addition to health risks there may be an issue on many sites with odours from vapours or trace gasses. In these cases the risk assessment should take account of the threshold concentration at which the vapour or gas can be detected. Many compounds have a very wide range of reported odour thresholds which can make the assessment difficult (e.g. for benzene reported detection values are in the range 0.78–160 ppm). However, the variations that are reported are probably caused in part by each individual's different olfactory sensitivity to these compounds. Therefore it is always wise to use a value from the lower end of the range to take account of the most sensitive people who are likely to be exposed to the vapour.

6.12 Tier 2 and 3 assessment of vapours and trace gases

The main part of this chapter has focused on risk assessment associated with the bulk ground gases (methane and carbon dioxide). These gases normally represent the worst case scenario (in terms of explosion and asphyxiation). The risk from trace gases and vapours (again in terms of explosion and asphyxiation) will be lower.

At present, there are no specific guidelines for assessing similar risks from trace gases and vapours. However, an initial assessment can normally be made using the same principles that have been applied to methane (i.e. consider 20% of the lower explosive limit as an initial screening value below which risks are considered acceptable).

Where trace gases and vapours are present there will also be a need to undertake a health risk assessment and this should follow the latest government and Environment Agency guidance using the Contaminated Land Exposure Assessment CLEA risk assessment model (or any other model that is considered appropriate, e.g. Risked Based Corective Action RBCA).

There is a wealth of information on the recommended approach that is available on the Environment Agency's website. The Scottish and Northern Ireland Forum for Environmental Research also provides a number of useful documents and spreadsheets on their website.

SUMMARY: Assessment of results

The analysis of data for a gassing site or site affected by gas migration should consider all available data together. It is not possible to interpret gas monitoring results without a desk study and ground investigation that identifies possible gas sources and migration pathways. The most common approach to assessing ground gas risk in the UK is described in CIRIA Report C659 and in NHBC guidance. This chapter provides guidance on using these documents.

The first step in any assessment of a site affected by ground gas or vapours is to construct a conceptual model. The desk study and site investigation data should be used to develop this. The first part of the gas risk assessment is to use the model to identify any source pathway and receptor linkages from the conceptual model. If there is no linkage then there is no risk. Once the source–pathway–receptor scenarios have been identified the risks associated with each one can be assessed qualitatively (i.e. descriptively).

When gas monitoring data becomes available this can be compared to the model. Comparison of the results of the gas monitoring with ground conditions can help identify what the main source of gas on a site is. For example it is often useful to group together the results of gas concentration or borehole flow rate for wells with response zones in a particular soil type. Graphical summaries and contour plots of data are useful to identify trends in data and to help identify if off- or on-site migration is occurring.

The risk assessment can be extended using the gas monitoring data and the site characterisation system from CIRIA Report C 659/665 or the NHBC traffic light system (for low-rise housing with an underfloor void).

A quantitative risk assessment is rarely required except on the most difficult or sensitive of sites. Sites that are being assessed under Part 2A of the Environmental Protection Act 1990 are likely to require a quantitative assessment to provide a robust indication that there is a pollutant linkage. The most commonly used method for quantitatively assessing risk on gassing sites uses a fault tree analysis to provide a numerical estimate of the risk. Guidance is provided about how to estimate some of the input parameters required in the fault tree analysis.

The justification of parameters and sensitivity analysis are an important part of fault tree analysis. Using fault tree analysis results to make statistical predictions about future system performance is complex and requires a great deal of experience. It is a complex and specialist field and anyone undertaking an analysis should be fully conversant with the application of the technique and be aware of how it is applied in other fields. It is important not to become unduly reliant on the numbers: remember they are only an aid to judgement.

Assessing the risk posed by gas to extensions to existing buildings is a difficult area. An approach is suggested.

CHAPTER SEVEN

Methods of gas protection

A wide range of gas protection measures is available. The most commonly used methods to protect developments can be divided into groups as follows:

- In ground vertical impermeable barriers used to prevent lateral gas migration (cut-off walls)
- In ground venting systems used to prevent lateral gas migration (vent barriers)
- In ground venting or active abstraction within gassing material to control gas pressures
- Impermeable barriers across building footprints (gas membranes)
- Ventilation below buildings (usually an underfloor void or venting layer). This can be active or passive
- Positive pressurisation below floor slabs (a particular form of active venting)
- Gas monitoring and alarm systems

The various methods are shown in Figure 7.1 which highlights those that are not applicable to low-rise housing (because they rely on regular maintenance).

There are also active gas pumping systems that are used in landfill sites to collect landfill gas for re-use or flaring. These are not commonly used on development sites and are not discussed further (sites where gas can be abstracted are usually not suitable for development).

Current UK construction practice adopts the concept of multiple gas protection measures, that comprise a gas control system, since no one protective measure is immune from factors unknown to or outside the control of the designer. Typically, protection measures increase in number and robustness with the higher characteristic situations defined in Chapter 6.

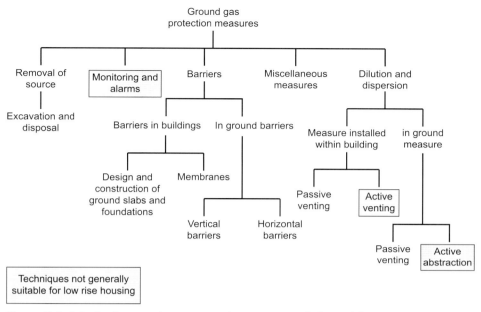

Figure 7.1 *Principal ground gas protection measures (adapted from Witherington and Boyle, 2007)*

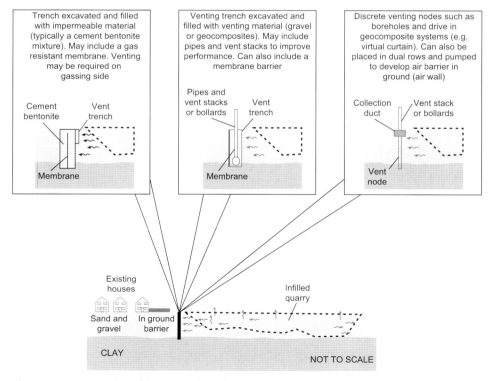

Figure 7.2 *Examples of in ground gas barrier and venting systems*

This chapter describes the various methods of protection that are available. Chapter 8 describes how to define the scope of protection for any site or building together with the calculations that are required to complete the detailed design of venting and pressurisation systems.

7.1 In ground vertical barriers or venting systems

A wide range of methods can be used to form in ground barriers. Some typical examples are shown in Figure 7.2. Comprehensive guidance on barrier design and construction has been given in: CIRIA Report C149 (Card, 1995), CIRIA Special Publication 124 (Privett *et al.*, 1996) and CIRIA Report C 557 (Barry *et al.*, 2001). The reader is referred to these documents for more detailed information.

All types of barrier or venting system should extend below the maximum depth that significant gas migration is considered likely to occur. In the majority of cases there is an impermeable stratum (such as the layer of clay in Figure 7.2) that provides a suitable lower impermeable layer the barrier or vent system can be toed into. In other cases the barrier or vent system is installed to 1 m below the lowest likely groundwater table. However, if groundwater is used as a barrier the risk of methane or other gases being dissolved in the groundwater should be assessed.

All impermeable barrier systems will require an assessment of the risk of uncontrolled gas migration along the length of the barrier and around the ends. If necessary, venting wells can be provided at the ends to manage this risk. The effect of barriers on groundwater flow also requires careful assessment.

A wide range of barriers and venting systems is available ranging from continuous impermeable barriers to venting systems with discrete venting nodes such as an array of boreholes. A summary of the most commonly used techniques is provided in the following pages.

Barrier systems that work on the principle of venting ground gas to the atmosphere should be designed to provide sufficient capacity to deal with the anticipated gas flows (vent trenches, boreholes, virtual curtain and air wall). The gas flow through the ground and within the vents can be calculated using Darcy's law and Fick's law (see Section 4.6). Sufficient ventilation should be provided to dilute the ground gas to safe levels at the vents by applying the principles which are used in passive underfloor ventilation design (see Chapter 8).

Technique	Description	Diagram/picture	Key advantages	Key disadvantages
Impermeable barrier Cement bentonite	1. Excavate trench and use bentonite slurry to support excavation 2. Tremie cement bentonite into bottom of excavation and recycle slurry 3. Insert membrane panel into cement bentonite if required 4. Cover surface of wall with protective clay or concrete layer		Can be constructed to depths in excess of 20 m Well-understood technology	Excavation and disposal of contaminated soils Disposal of bentonite slurry Can be breached by unplanned excavations by utilities
Impermeable barrier Clay filled trench	1. Excavate trench 2. Backfill with clay and compact 3. Finished surface may require protection		Well-understood technology Easily installed by groundworkers	Depth limited to about 5 m Excavation and disposal of contaminated soils If ground is unstable will require supporting or battering of excavation sides Can be breached by unplanned excavations by utilities

Technique	Description	Diagram/picture	Key advantages	Key disadvantages
Impermeable barrier / Membrane in trench	1. Excavate trench 2. Place impermeable membrane 3. Backfill: can use as-dug material if suitable		Well-understood technology Membranes can resist a wide range of contaminants	Depth limited to about 5 m Excavation and disposal of contaminated soils If ground is unstable will require supporting or battering of excavation sides Can be breached by unplanned excavations by utilities Membrane can be damaged during backfilling if it is not adequately specified and protected
Impermeable barrier / Sealed sheet pile wall	Drive in sheet piles with clutches sealed to prevent gas migration. Sealing is achieved by using proprietary mastic or silicone sealants.		Can be installed to depths of about 15 m Limited space required No excavation of contaminated soils Can act as retaining wall	Vibration can be an issue (but push-in systems are available) Obstructions can prevent design depth being reached Quality of sealing can be difficult to control

Technique	Description	Diagram/picture	Key advantages	Key disadvantages
Venting trench Gravel or geocomposite	1. Excavate trench 2. Backfill with permeable gravel (typically 20 mm pea gravel). 3. Membranes can also be included to provide additional protection. 4. Additional air flow can be provided by using pipes within the gravel connected to vertical vents.		Well-understood technology Flexible Easily installed	Depth limited to about 5 m Excavation and disposal of contaminated soils If ground is unstable will require supporting or battering of excavation sides Can be breached by unplanned excavations by utilities
Venting wells	Large diameter boreholes with well casing and gravel surround installed Can also be installed as stone columns using ground improvement plant Ventilation via stacks, bollards or ground level boxes		Well-understood technology Flexible layout Easily installed Can be installed to about 30 m depth Easily installed around services	Need to be at close centres to be effective Excavation and disposal of contaminated soil (although low displacement methods can be used) Obstructions can prevent design depth being reached

Technique	Description	Diagram/picture	Key advantages	Key disadvantages
Virtual curtain passive venting system	1. Excavate shallow (1 m deep) trench for collection duct 2. Vibro insert steel mandrel and insert geocomposite nodes into it. Withdraw steel mandrel leaving geocomposite in ground 3. Install collection duct over nodes 4. Backfill trench and connect to vent stacks or bollards		Flexible Minimal excavation and disposal of contaminated soils Quick installation compared to vent trenches and impermeable barriers in trenches	Maximum depth of installation of 10 m Vibration can be a problem but this is rare Obstructions can prevent design depth being reached Difficult to install into bagged domestic refuse
Airwall active air flow barrier	1. Install two rows of boreholes 2. Connect up pipework and pumps to boreholes 3. Set up pumps and controls to pump air into one row of wells and suck air out of the other (this occurs in response to gas monitors in the wells detecting elevated gas concentrations) 4. Flow of gas in ground in opposite direction to gas migration provides barrier to gas migration		Flexible Minimal excavation of contaminated soils Can be installed to 30 m depth Zone of influence below toe of boreholes	Ongoing regular maintenance and running cost requirements Local authorities may be reluctant to accept active gas protection on residential developments

Monitoring wells can be installed on both sides of a barrier to demonstrate that gas migration has been prevented or reduced. Care must be taken when installing such monitoring wells that they only intercept the gas migration pathway and do not collect gas from other sources (e.g. shallow made ground) which could wrongly suggest that gas migration is still occurring after the installation of a barrier (Figure 7.3).

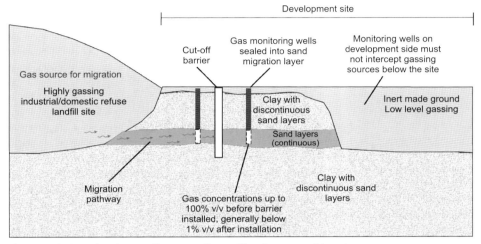

Figure 7.3 *Monitoring wells to confirm effectiveness of barrier*

7.2 Building protection

The main methods of protecting buildings from gas ingress are:

- Provision of a robust floor slab and gas-resistant barrier across the floor slab combined with
- Underfloor passive venting or positive pressurisation

Many urban developments include basement car parks that are well ventilated to deal with exhaust gases from cars. They are often constructed with thick reinforced concrete retaining walls and cast in situ suspended floor slabs. The basement is in effect a very large and effective ventilated void that provides more than adequate protection against low levels of ground gas. Care should be taken, however, to ensure that features such as lift shafts or service risers which extend to the base of the car park are provided with adequate protection.

For commercial developments on more highly gassing sites additional protection may be provided by using the following:

- Active ventilation as a backup to passive systems
- Monitoring and alarm systems

Clay capping layers may also be used as additional protection layers to minimise the risk of vertical migration. However, these are difficult to seal around
foundations that penetrate them (e.g. piles) and should not be used as a replacement for either a ventilated void below the floor slab or to replace membranes
across the floor slab.

7.3 Floor slabs and gas-resistant membranes

7.3.1 Floor slabs

The floor slab is the first barrier against gas ingress. There are several types of
floor slab that are commonly seen on developments in the UK. Each type will
offer a degree of protection against gas ingress into the finished structure.

Reinforced concrete cast in situ suspended floor slabs can give reasonable gas
resistance, if they are detailed correctly with anti-crack reinforcement. They
also need to be sealed at joints (possibly with water bars: a design measure to
ensure joints are watertight) and sealed at all service penetrations with water
bars. Evidence from Building Research Establishment (BRE) research has shown
that reinforced concrete rafts with a venting layer below give the best level of
protection to gas ingress.

Block and beam slabs used in many housing developments have many joints
between the blocks and beams which means they provide minimal resistance to
ground gas ingress. However, a well-constructed block and beam floor should
have a sand:cement grout brushed into the surface to fill all the joints which
can improve gas resistance to a limited extent. This also gives a better surface
into which to lay a gas-resistant membrane (see Figure 7.4).

If a ground-bearing slab is used on a site affected by ground gas it must be
designed to minimise the risk of differential movement between it and the walls
of the building. Such differential movement has been one of the main causes of
damage to gas-resistant membranes (the movement has torn the membrane).
The membrane must therefore be specified and detailed to accommodate any
likely movement (e.g. by providing a fold in the membrane that allows movement, by limiting the movement of the slab or by specifying membranes with a
high elongation before failure).

7.3.2 Gas-resistant membranes

The gas resistance of floor slabs is improved by providing gas-resistant
membranes. The purpose of the gas-resistant membrane is to provide a low
permeability barrier against the ingress of gas from the ground into the building
fabric.

Figure 7.4 *Well-constructed block and beam floor slab ready to receive a membrane*

There are many different types of membrane on the market each with advantages and disadvantages and with varying degrees of methane permeability. However, no membrane is completely impermeable to ground gas or vapours. The permeability to methane of some typical proprietary membranes is given in Table 7.1, together with the time for gas to leak through and fill the overlying room with methane to 5% v/v. The results show that preventing holes due to damage and incorrect installation is the critical issue, as once any of the membranes have even the slightest puncture, the rate of gas flow into an overlying room will increase by a factor of over 12 million times.

It is recommended that gas membranes are specified on the basis of the need to survive construction, i.e. use one that is resistant topuncture and tear. The joints should be sealed in accordance with the manufacturer's instructions. Thickness should not be the main method of determining the suitability of a membrane for a particular site, although it can indirectly influence some of the key performance parameters.

For guidance, it should be clear that proprietary gas-resistant membranes are manufactured using robust materials that are often reinforced and have thicknesses from 1600 g (0.4 mm) upwards. DPMs are usually made from lower quality materials (typically lower grade low density polyethylene(LDPE) than gas-resistant membranes and thus require a greater thickness. A 1200 g polyethylene (PE) DPM material (0.3 mm) as shown in previous BRE guidance is unlikely to be installed correctly (it will usually be installed by general

Table 7.1 *Methane permeability of membranes*

Membrane	Methane permeability (ml/m²/24h/atm)	Time to fill room to 5% v/v methane through intact membrane (yr)	Time to fill room to 5% v/v methane through punctured membrane: 1 mm diameter puncture (h)
Polypropylene (PP) 1 mm thick (4000 gauge)	0.15	3.7 million	12
Polyethylene (PE) 0.4 mm thick (1600 gauge) reinforced	31	17,675	12
Low density polyethylene (LDPE) sandwich with aluminium foil core. 0.4 mm thick (1600 gauge) reinforced	<0.001	547 million	12

Notes
1. Room assumed to be 2 m high × 5 m × 4 m
2. Driving pressure of gas through membrane is 50 Pa
3. Divide gauge by four to establish equivalent thickness in microns

labourers) and is easily damaged during installation and by follow on trades. Therefore the minimum thickness for a DPM (as opposed to a proprietary gas-resistant membrane) used as a gas protection is 2000 g. In any event the BRE guidance suggests that the use of 1200 g PE is only suitable for sites where the methane concentrations are less than 1%. In very low risk situations (Characteristic Situation 2 with an in situ reinforced concrete floor slab) a 1200 g DPM may be acceptable provided it is provided with adequate protection and is verified as adequate by extensive independent inspection (see Chapter 9).

Comparison of the performance characteristics of different membranes can be difficult because different test methods are sometimes used to measure the same property. In the UK membrane testing should be undertaken following the methods in British or European standards (BS or EN). It is common to see test data using American standards (American Society for Testing and Materials (ASTM)) but these results are not directly comparable with data from British or European standard tests. The only exception to this is where a standard is an ISO (International Standards Organization) standard which is recognised internationally.

When specifying a membrane only those criteria that have an impact on the performance should be stated. The quoted limits should reflect the necessary level of performance that is required in a particular application. The critical properties that should be specified for a gas-resistant membrane are summarised in Table 7.2 and focus on the short-term survivability during installation.

Table 7.2 *Suggested membrane testing*

Characteristic	Test method	Unit	Considerations
Tensile properties	BS EN ISO 10319: 1996 Geotextiles. Wide-width tensile test	kN/m	The greater the tensile strength the greater the resistance to tearing and puncture. Amount of elongation before failure is important where movement of the membrane is anticipated
CBR puncture resistance	BS EN ISO 12236: 2006 Geosynthetics: static puncture test (CBR test)	N	Acceptable puncture resistance depends factors such as the nature of the surface onto which the membrane is laid (e.g. is it smooth concrete or a jagged aggregate?), type of work that will be carried out on it after installation (e.g. will concrete trucks need to run on it?), type of protection provided (boards, geotextile fleece) and level of independent inspection. The higher the risk of damage occurring the greater the CBR puncture resistance that should be specified. Example values for HDPE membrane: 160 N for 0.5 m thick sheet and 320 N for a 1 mm thick sheet
Resistance to tearing (nail shank)	BS EN 12310-2: 2000 Flexible sheets for waterproofing. Determination of resistance to tearing (nail shank). Plastic and rubber sheets for roof waterproofing	N	Acceptable level of resistance to tearing depends on the same factors as CBR puncture resistance
Methane permeability	BS903: part A30 (modified)	ml/m²/ day	Minimum value should be based on a risk assessment and time for gas to migrate through membrane and accumulate to a hazardous concentration. Typically a minimum value of 35 (ml/m²/24 h/atm) is more than adequate

Membranes should ideally be resistant to puncture abrasion and tearing, in addition to ultraviolet (UV) light, shrinkage, water, organic solvents and bacteriological action. In general terms the thinner the material, the more susceptible the membrane will be to mechanical or chemical damage.

In practice, because of the nature of construction sites, the durability, survivability and robustness of membranes are more significant properties than methane (or any other gas) permeability. In addition the need for high quality workmanship during the installation of the gas-resistant membrane should not be under-estimated. The health and safety of the occupants of the building depends on its satisfactory performance. If installed incorrectly or damaged during the construction process, the membrane is rendered ineffective and therefore fails to provide adequate protection against the ingress of ground gas or vapours. This is why independent inspection of all membranes is a vital part of the validation process (see Chapter 9).

A lower specification membrane (e.g. 2000 g DPM) that is well installed and not damaged by follow on trades is far better than a poorly installed high quality membrane that is extensively damaged or not correctly sealed. However, it should be noted that it is usually more difficult to seal lower quality membranes (e.g. they cannot be easily welded) and they are more prone to damage and so require more protection than high quality robust membranes.

7.3.3 Membrane types

PE and polypropylene (PP) are the principal materials used in the manufacture of gas-resistant membranes. PE is available in a variety of thicknesses and types:

- HDPE (high density polyethylene)
- LDPE (low density polyethylene (this can be low-grade DPM or higher quality reinforced membrane))
- MDPE (medium density polyethylene) which is not commonly used as a gas-resistant membrane material
- LLDPE (linear low density polyethylene)

The membranes can also be reinforced to improve the durability of the material and prevent over-elongation of the membrane. Other types of materials and membranes are available and a specific assessment should be made where these materials are proposed (e.g. liquid applied membranes and aluminium bonded to geotextile membranes). The disadvantages and points to consider for various types of membrane are provided in Table 7.3.

There is a common misconception that carbon dioxide always poses a lower risk than methane and that somehow a lower specification membrane can be used. This has probably arisen because slightly elevated carbon dioxide concentrations on their own are often associated with low generation sources. Thus, it is the overall risk due to the source that is low. It is not simply because only carbon dioxide is present. There are situations where carbon dioxide can pose a significant risk and it is up to the consultants and other designers to assess this risk and the construction conditions and to propose a suitable membrane. It is not adequate to simply propose a 'carbon dioxide' membrane without a full justification as to why the risk is lower. If it is considered that carbon dioxide can migrate into a building then a suitably robust membrane should be provided.

Different membranes to those for methane and carbon dioxide are rarely required to prevent vapour migration into buildings from hydrocarbons. Any of the thermoplastic materials listed above will be sufficiently resistant to all of the most commonly encountered gases/vapours. Quality of installation is the key issue (DETR, 1997).

Table 7.3 *Gas-resistant membranes*

Advantages	Disadvantages	Points to consider
PP		
Excellent chemical resistance	Relatively high methane permeability	PP requires specialist welding and should not be tape jointed. Usually supplied in large rolls and therefore is more suitable to large floor areas and also requires plant for handling on site. Welds can be pressure tested as part of quality assurance (QA) to ascertain joint integrity.
Good elongation before failure	Higher unit cost	
	Reduced adhesion with self-adhesive membranes	
Flexible		
Does not suffer stress cracking		Flexible so more suitable to complex details and high elongation before rupture so suitable where settlement may occur
HDPE		
Excellent chemical resistance	Can be prone to stress cracking	HDPE requires specialist welding and should not be tape jointed. Usually supplied in large rolls and therefore is more suitable to large floor areas and also requires plant for handling on site. Welds can be pressure tested as part of QA to ascertain joint integrity. Complex detailing can be hard to achieve due to the rigidity of the membrane
Good puncture resistance	Rigid	
	Difficult to achieve complex detailing	
Large roll sizes, therefore less jointing		
Low unit cost	Heavy and difficult to handle	
	Reduced adhesion with self-adhesive membranes	

Table 7.3 *Gas-resistant membranes (continued)*

Advantages	Disadvantages	Points to consider
LDPE DPM		
Good chemical resistance Flexible Low unit cost Easy handling Widely available	Low-grade recycled raw materials Easily damaged Relatively high methane permeability Low UV resistance	Most standard PE DPM is manufactured from recycled LDPE building film which is considered to be a low-grade LDPE. Tolerances for building film thickness can be as much as ±10%. Minimum thickness of this type of membrane should be 0.5 mm or 2000 g (CIRIA,006)
Reinforced LDPE or HDPE with an aluminium core		
Best resistance to ground gas or vapour migration if correctly installed. Flexible Low cost Increased tear resistance Good chemical resistance	Aluminium core can corrode on thinner PP versions from prolonged contact with alkaline substances. Low UV resistance Foils susceptible to rupture where settlement occurs, unless allowed for in design and installation	Reinforced LDPE membrane is normally manufactured from high quality virgin polymer laminating film as the thickness tolerance is critical for the heat bonding process (this should be confirmed). If this is not the case the reinforced membrane will be prone to delamination. Minimum recommended thickness 0.4 mm to provide sufficient protection to aluminium from alkaline substances. PP-type membranes bonded to aluminium are not acceptable due to poor adhesion of PP to aluminium

The present authors are of the opinion that the use of geosynthetic clay liners (a thin layer of dry clay powder sandwiched between two geotextiles) as gas-resistant membranes below buildings is not acceptable. This is because the liner relies on the bentonite material becoming wet to form a barrier and this cannot be guaranteed. Even if it is pre-wetted the clay can dry out and crack, which will allow gas to migrate through it.

Whichever type of membrane is used the installation should be inspected by an independent party after it has been laid and before it is covered over to ensure that it is correctly installed (see Chapter 9).

Where membranes are used in the ground as a vertical barrier they can be subject to high stresses and puncture forces from the surrounding ground, settlement of backfill causing drag down etc. Very robust materials and protection must therefore be provided (see Card, 1995; Privett *et al.*, 1996).

7.3.4 Gas membrane manufacture

Gas-resistant membranes should be manufactured under a quality system independently certificated to BS EN ISO 9001 or equivalent. The manufacturer should carry out regular in house quality control tests and inspections to ensure that the finished products meet/exceed quoted material specification/ data sheets and these should be available on request. Consideration should be given to independent testing and or inspection of samples before approval for use is granted. It is also useful to carry out independent inspection of random samples taken from the materials delivered to site. Traceable quality control records must be maintained by the manufacturer with all rolls individually labelled with a batch number. If any problems with manufacturing quality arise on site during installation this allows the extent of the problem to be identified.

7.3.5 Sealing

Sealing and jointing membranes can be achieved using welding or specialist adhesive tapes. For both operations the two most critical factors are:

- Membrane must be clean and dry to allow welding or taping
- Temperature must not be too cold to allow welding or taping

At cold temperatures it may not be possible to joint membranes or pre-warming using hot air may be required (typically at temperatures less than 5°C). It is also essential that there is sufficient membrane material to form a weld. More information on the methods of jointing has been given in CIRIA Special Publication 125 (Privett *et al.*, 1996).

For taped joints the tape used should be that recommended by the manufacturer. Some of the issues that need to be considered when using taped joints are:

- Compatibility: numerous tapes are not actually adhesive to HDPE and PP
- Age hardening: many polymerised bitumen tapes (the majority of adhesive tapes) age harden and become less adhesive, stiffen and can eventually de-bond. Although this may be less of a problem in horizontal installations with a weight of concrete on them, it does mean that vertical joints require mechanical support in the form of battens otherwise the joints can peel open
- Adhesive tapes usually fail at the point of installation due to dust contamination: either dust on the membrane, dust from the substrate or dust on the adhesive tape
- Effectiveness of tape joints relies on a firm pressure being applied to the overlaps and for this reason (and the previous dust concern)

reputable manufacturers of adhesive membranes and tapes all recommend a concrete blinding to provide a supporting substrate

- Tape jointed systems are more susceptible to moisture at the time of installation than welded joints
- Tensile strength and adhesion on taped joints are far lower than on welded joints. They are prone to de-bonding during installation as a result of thermal expansion if the membrane is subject to wide temperature gradients
- Tape joints cannot be tested as comprehensively as welded joints, as they rely on the 'screwdriver test' and visual inspection

7.3.6 Detailing

Of equal importance to the performance of a membrane is good detailing of joints etc., especially around complex structural forms (see Figure 7.5). Good detailing should ensure that the installation and jointing of the membrane is as easy as possible. This should be considered at the design stage to prevent problems during installation. An experienced engineer on site during the installation can also be beneficial in dealing with detailing problems.

Figure 7.5 *Complex structural forms require good detailing and careful installation of membranes to ensure they are gas tight*

It should be noted that if there is likely to be any differential movement between the floor slab and the wall foundations this must be allowed for in the design and specification of the gas membrane.

7.3.7 Protection

Membranes can be provided with extra protection from damage using proprietary protection boards or thick geotextiles. This is similar to the approach used in landfill lining.

7.3.8 Durability

Membranes will degenerate over time due to processes such as oxidation, biological degradation, chemical attack and degradation caused by UV light. It is necessary to understand how long a given membrane will maintain its performance over the design life. As a general rule membranes used in gas protection are not exposed to UV light after installation. The membranes should only be exposed to UV for a short period of time during installation. In addition, membranes usually have UV stabilisers added to protect against attack by UV. Where a membrane is exposed to UV light its life will be reduced.

Another common issue is the resistance of membranes to attack by hydrocarbon vapours. This is unlikely on most sites as they are usually above a venting layer or above the floor slab and are not in direct contact with ground contamination. The vapours passing through the system will be at very low concentrations that are unlikely to cause damage to the materials used to make gas membranes (most testing that shows damage by hydrocarbons to membranes is using 100% concentration liquids).

Oxidation and biological degradation will occur, but for most membranes stabilisers are added to the materials and this will be a very slow process. The main effect of the aging processes is to reduce the strength of the membrane. Where the liner is subjected to long-term stresses, stress cracking will also lead to the development of holes. As gas membranes do not usually have to carry any great load after installation the adverse effects associated with these aging processes should therefore not be significant. Gas-resistant membranes should in most cases therefore have a very long service life (at least 100 years or more) at an ambient temperature of about 20°C. The service life will be reduced as temperature increases and where membranes are subject to long-term stress.

It has also been known for rodents to gnaw through exposed membranes, although the conditions that would allow this (an exposed membrane) are uncommon. Where membranes are used in vertical in ground barriers they must be specified to resist root penetration if close to trees.

7.4 Passive ventilation below buildings

Passive ventilation below buildings is designed to use wind effects on the sides of buildings to ventilate an underfloor void. It is also known as natural ventilation and with the increasing emphasis on energy efficiency in buildings it is the preferred option wherever possible (as opposed to systems that use active fans). Many contaminated land officers are reluctant to accept proposals for active systems because of the long-term management and maintenance requirements. The wind develops pressure and suction (see Figure 7.6) that drives fresh air through the void, thus diluting the gas emissions so they can be safely dispersed to the atmosphere.

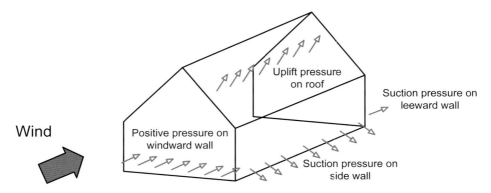

Figure 7.6 *Wind-induced pressure and suction*

This section discusses the various methods of forming an underfloor void. Advice on how to complete the design calculations to ensure that there is sufficient ventilation is provided in Chapter 8.

Underfloor voids can take a number of different forms:

- An open void below a pre-cast suspended slab
- A basement car park
- A void former below a cast in situ floor slab (which can be ground bearing or designed as suspended). The void former simply supports the wet concrete until it is set, and leaves a void below the slab. This is a common method of construction in warehouses or industrial buildings

7.4.1 Open void below a pre-cast suspended slab

This type of construction is frequently used in housing. Typically a block and beam floor is used (see Figure 7.4). The pre-cast floors allow an open void below that is typically 150 mm high. There will be sleeper walls below most buildings that need to be vented when this type of construction is used (see Chapter 8). This is the most efficient type of underfloor ventilation. In addition the use of suspended floor slabs reduces the risk of damage to a gas-resistant membrane caused by differential settlement between a ground bearing slab and the walls of a building.

7.4.2 Basements

Modern basement design requires efficient air ventilation to maintain indoor air quality and prevent damp. Table 1 of BS8102: 1990 defines four grades of basements (types 1–4) depending on their intended use. For the lower grades of basement (types 1 and 2), typically assigned for underground car parks, the minimum ventilation requirement is designed to vent fumes from car exhausts and prevent them building up to unacceptable concentrations and also to remove condensation. These fumes include carbon monoxide and the volumes that are emitted directly into the car parking space are usually far greater than the volumes of gas likely to migrate from the ground. Therefore the ventilation provided to deal with car exhausts is usually more than adequate to also deal with ground gas or vapour emissions and the car park is, in effect, a very large ventilated void.

However, care needs to be taken in detailing basement car parks as there may be places, such as bin stores or parts of the main building that are not over the car park, that may require some form of specific ground gas protection.

Higher grades of basement (types 3 and 4), usually assigned for habitable spaces, must be designed to prevent water ingress. This is usually achieved by applying a tanking membrane to either the external or internal walls and ground floor of the basement. The waterproofing membrane usually also acts as a gas-resistant membrane. Further information on waterproofing basements has been provided in CIRIA Report 139 (Johnson, 1995).

It should also be borne in mind that the basement construction on many sites may remove most if not all of the gassing source and thus the risk from ground gas is reduced or removed.

7.4.3 Void formers

There are several different types of void former in use but the main ones are either polystyrene, geocomposites or pipes and open graded gravel.

A polystyrene void former comprises a slab of polystyrene with legs or stools that support it off the ground. It provides a level surface with a void below. A membrane is placed over it before casting the floor slab. The system is available with a variety of void heights and its performance has been demonstrated in the Partners in Technology report (DETR, 1997). There are also different grades and shapes for differing load bearing capacities. An example is shown in Figure 7.7 where the legs or stools are clearly visible on the lower part of the void former. A typical installation detail is also shown in Figure 7.7. The void below the polystyrene is vented using either air bricks, stacks or bollards.

The ventilation and air flow through the void can be estimated in the same way as for geocomposite systems (see Chapter 8) using the intrinsic permeability quoted in Table 7.2.

The Partners in Technology report indicated that polystyrene shuttering is almost as effective as an open void of similar depth. The effectiveness is, as would be expected, sensitive to the actual venting provided. The modelling demonstrated that the polystyrene systems can be designed to vent below 30 m

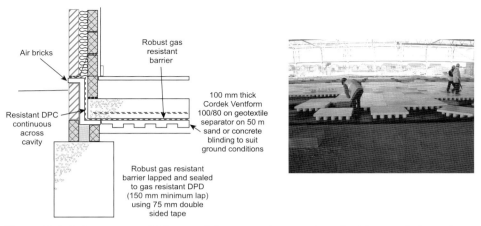

Air bricks

Robust gas
resistant
barrier

100 mm thick
Cordek Ventform
100/80 on geotextile
separator on 50 m
sand or concrete
blinding to suit
ground conditions

Resistant DPC
continuous
across
cavity

Robust gas resistant
barrier lapped and sealed
to gas resistant DPD
(150 mm minimum lap)
using 75 mm double
sided tape

Figure 7.7 *Polystyrene void former (photograph reproduced with permission of Cordek Ltd., Horsham, UK)*

wide buildings for all but the very worst gas regimes, using low-level vents. If high-level vents are used and the vent area increased it is possible to use these systems on much wider buildings for lower gas regimes.

Hydrocarbons can adversely affect polystyrene, but as with membranes, the concentration of vapours in a well-designed venting system should be very low and should not normally affect the polystyrene significantly. Even if it was adversely affected, once a suspended slab is constructed it does not rely on the polystyrene for any structural performance and thus the impact of any deterioration in the polystyrene will be minimal. For slabs that rely on the polystyrene for permanent support the risk of potential deterioration should be considered.

Many types and thicknesses of geocomposite are used as void formers in gas protection systems. One of the most common has a cuspated core that has downstands on it. A geotextile is bonded to the end of the cuspates and is laid on the ground with the geotextile at the bottom. Others include geotextile fleeces with small diameter tubes running through and geonets with geotextile layers bonded to either side. A typical cuspated core geocomposite is shown in Figure 7.8 (upside down) together with a typical installation detail.

Careful detailing of geocomposite systems is required to ensure that ventilation is provided to all areas below floor slabs where there are obstructions such as downstand beams, foundations and pile caps. Pipes are often connected to the geocomposite to continue the flow path either through or around the obstruction.

The Partners in Technology report (DETR, 1997) suggested that geocomposites are not especially effective below buildings that are greater than 30 m wide. This is because they assume venting via low air bricks and thus air flow is

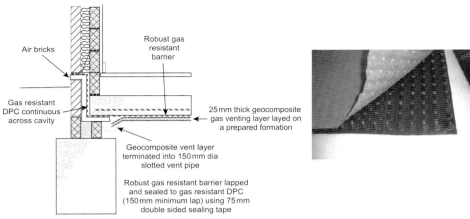

Figure 7.8 *Typical geocomposite void former*

limited. It is possible to design effective geocomposite systems for wider buildings but high-level stacks and a high ventilation area are needed at the side to generate sufficient air flow through them. Active venting can also be provided as a backup to the passive ventilation in wider buildings.

With active systems where greater air flow can be achieved thinner geocomposites can be used. Typical air transmissivity rates and intrinsic permeabilities for a range of geocomposites are given in Table 7.4.

The considerations regarding the effects of hydrocarbons in polystyrene void formers also apply to geocomposites.

Table 7.4 *Typical air flow rates and permeability of void formers*

Type/thickness	Estimated air transmittivity rate at 5.4 Pa pressure[A] (l/s/m width)	Typical intrinsic permeability (m²)[B]
Geocomposite 40 mm double-sided cuspate waffle core	1.2 (Lower than 25 mm flatback products because of the tortuous flow path formed by the core)	2.0×10^{-6}
Geocomposite 25 mm flatback single-sided cuspate columnar core	4.5	1.2×10^{-5}
Geocomposite 12 mm flatback single-sided cuspate columnar core	0.6	3.4×10^{-6}
Geocomposite 6 mm flatback single-sided cuspate columnar core	0.09	9.7×10^{-7} [C]

Table 7.4 *Typical air flow rates and permeability of void formers (continued)*

Type/thickness	Estimated air transmittivity rate at 5.4 Pa pressure[A] (l/s/m width)	Typical intrinsic permeability (m^2)[B]
Type 1 subbase (and similar materials)	0.06	2.7×10^{-8}
20 mm open graded gravel	0.2	9.9×10^{-8}
Polystyrene void former: Ventform 80	24.5	7.5×10^{-5}
Polystyrene void former: Ventform 100	53.9	1.1×10^{-4}
Polystyrene void former: Ventform 150	329	3.7×10^{-4}
Polystyrene void former: Ventform 200	743	5.0×10^{-4}

[A] Calculated using intrinsic permeability and a pressure of 5.4 Pa over 20 m length
[B] Determined from computational fluid dynamics modelling as in Partners in Technology Report (DETR, 1997) unless stated otherwise
[C] Estimated pro rata from 25 mm and 12 mm products

Gravel and pipes or geocomposite strips in gravel can also be used as void formers on sites where the ground gas poses a low risk (see Figure 7.9). The pipes or geocomposites are laid within a layer of gravel below the floor slab to provide additional ventilation capacity. However, the permeability of the gravel is critical to the performance of the system and a relatively coarse, single-size material with minimum fines is the most effective (e.g. 4/20 or 4/40 material). Type 1 subbase or similar material with in excess of 5% fines (material with a particle size less than 63 µm) is unlikely to provide an effective ventilation layer. The open area (or slot area) of a pipe is also critical. The Partners in Technology Report (DETR, 1997) recommended an open area of at least 10–20% in order to assume that there is no resistance to gas entry into the pipes. Below this area, the entry of gas into the pipe will become the limiting factor in the design. It should be noted that a normal land drainage pipe is unlikely to meet this requirement. Where pipes or geocomposites are used in gravel they should be interleaved to prevent short-circuiting of air flow through the pipes or composite (see Figure 7.10).

The granular layer is often used as a general construction or piling platform before the floor slab is constructed. This usually results in clogging of the venting layer and is not recommended, unless a sacrificial surface layer is removed before placement of the geocomposite or pipes.

Figure 7.9 *Geocomposite strips in gravel*

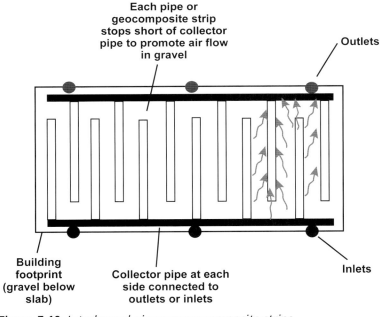

Figure 7.10 *Interleaved pipes or geocomposite strips*

Careful detailing of pipe or geocomposite strip systems is required to ensure that ventilation is provided to all areas below floor slabs in between obstructions such as downstand beams and foundations. The maximum distance a pipe or strip should be from the edge of the building is 1 m (DETR, 1997).

7.4.4 Venting components

A wide variety of venting components is available for use with passive gas protection systems (see Figure 7.11). The specific type of vent depends on a number of factors such as: cost, aesthetics, location and venting efficiency. The venting efficiency of the outlet components is predominantly governed by their height above ground level and ventilation area. Rotating cowls can be added to vent stacks to increase the ventilation performance.

Figure 7.11 *Different types of venting components*

It is possible to obtain custom vents to suit the requirements of any particular project. They can also be disguised as architectural features.

7.5 Active venting below buildings

Active venting uses fans to enhance the airflow through underfloor voids. The number of fans provided depends on the required air flow. It is best if fans that allow passive air flow when they are not operating are used and that a gas monitoring system is installed in the underfloor void. The monitoring system activates the fans to increase air flow when the gas levels exceed set values.

Alternatively, they can be designed to run continuously at low speeds (similar to positive pressurisation systems (see Section 7.6)). There is, however, a common misconception (that is unfounded) that fans are more reliable if they run continuously. This is incorrect as fans will have finite run lifetime (that should be known by the manufacturer) and the more they run the shorter the time before a replacement is required. An inline fan system that acts as a passive system most of the time is shown in Figure 7.12.

Figure 7.12 *Active venting system*

All the electronic components and fans used in any type of active venting or pressurisation system require regular maintenance, although this is relatively easy. However, the likelihood of a system not being maintained has to be seriously considered and measures put in place to prevent the system being neglected. For this reason it is generally accepted that active systems should only be used for gas protection if it has been shown that a passive system is not sufficient (this is similar to the guidance for basement car parks given by the Institution of Structural Engineers (2002)). They are not generally suitable for housing and may also conflict with the requirements for energy efficiency in buildings.

7.6 Positive pressurisation below buildings

Positive pressurisation is a particular form of active system that works by pumping air into the void or void former below a floor slab to create a zone of pressure that is greater than the pressure of the gas in the ground. This prevents gas migrating to below the floor slab (see Figure 7.13, colour section). Many of the comments regarding maintenance and fan life also apply to these systems.

The air flow into the void must be sufficient to maintain a pressure that is greater than the pressure driving the gas from the ground. For a simple void the limiting factor is likely to be cracks and gaps in the foundation construction that allow air to escape and reduce the pressure below the slab. For other systems the permeability of the ground will have a critical influence on the performance of the system.

There is some concern with these systems that if there is any defect in the gas membrane then the positive pressure means gas may be driven up into building. Thus the specification of the gas membrane is more critical with positive pressure systems and a robust well-installed membrane is vital.

These systems also act as an impermeable area and thus prevent gas escaping from the ground under a building. This can force gas to migrate laterally in the ground (the positive pressure can enhance this effect). Thus a perimeter collection system that extends below the depth of influence of the pressure blanket is required to mitigate these effects.

Guidance from the Californian Department of Toxic Substances Control (State of California, 2005) states that the following need to be considered when actively pumping air into voids below floor slabs:

- The system should be designed so that it does not create increased pressures under the building that may force methane into the building or into unprotected neighbouring properties or structures. Sensors to monitor subslab pressures should be considered
- Utility trench dams should be included to prevent methane from being forced into utility trenches, pavement subgrade and/or other conduits

As with other active systems there is a greater ongoing maintenance requirement than for passive systems and the implications of this not being undertaken should be carefully assessed. Active systems are often observed to have ceased working after a few years because the maintenance contracts have lapsed. There is also an issue of energy efficiency with all active systems. The Building Regulations Part L require buildings to use the least amount of energy possible and passive or natural systems are preferred in this respect.

7.7 Gas monitoring and alarm systems

Gas monitoring and alarms are sometimes used to give added assurance that other methods of protection are working correctly. They are also used to retrofit into existing buildings where it may be difficult to install other types of protection, or where emergency protection is required to be installed quickly before other methods can be installed. Because of the need for regular maintenance it is not appropriate in housing, except as an emergency measure in existing developments.

The best place to use gas monitoring is in an underfloor void so that advance warning of gas migration is given before it builds up in occupied spaces.

The most common type of system currently used to monitor methane and carbon dioxide in buildings is an infrared sensor. The sensor is located at a central position (see Figure 7.14) and each monitoring point around a building is connected to the sensor by a nylon sample tube. Air is sampled from each point at regular intervals by pumping it up the sample line to the central sensor unit. This is usually achieved using a series of solenoid valves connected to a shared pump. Depending on the type of system a single control/sensor unit will monitor 8–40 locations. These systems are reliable and easy to maintain.

Where sensors are located in the occupied building space they should be positioned at points where gas ingress is most likely, for example:

- Around service entries
- In small rooms or cupboards with little natural ventilation
- Close to wall/floor joints
- In rooms that contain ignition sources such as boilers or electrical plant rooms

Figure 7.14 *Alarm panel in a plant room*

The systems typically monitor for methane (concentration range of 0–100% lower explosive limit (LEL)) and carbon dioxide (concentration range of 0–5% by volume in air) but this will always be site specific and should take account of the specific gases present in the ground below a site.

The alarm sequence for a monitoring system is based on site specific conditions, but a typical protocol would be:

- Underfloor void, low-level alarm: 20% LEL of methane
- Underfloor void. high-level alarm: 40% LEL of methane
- Internal monitoring within the building, low-level alarm: 10% LEL of methane
- Internal monitoring within the building, high-level alarm: 20% LEL of methane
- Internal monitoring within the building, low-level alarm: 0.5% carbon dioxide by volume in air
- Internal monitoring within the building, high-level alarm: 1.5% carbon dioxide by volume in air

Monitoring systems are often used in conjunction with active systems. When gas is detected in the underfloor voids the ventilation is automatically increased to reduce the gas concentration. For the occupied spaces procedures need to be put in place to respond to alarms that indicate elevated gas concentrations within the building. The low-level alarm may require investigation of the cause and ventilation of the building. At the high-level alarm the normal procedure is to evacuate and ventilate buildings.

Detection and alarm systems are designed to operate continuously with service intervals of between six months and one year.

The main sensor controls are typically located in the plant room or security office, but any alarms should be located on mimic panels where they can be clearly seen and acted on by the users of the building. The system should allow recall of alarm events, affected zone, gas concentrations, date and time. There should be a clear indication of the fault zone on the display. The system should be powered by an uninterrupted power supply (UPS) so that it continues to operate even during power cuts. A UPS is a set of batteries and a control unit that senses when electrical power is cut and uses the battery backup to power the system.

Remote (or mimic) panels are required with gas detection and alarm systems. These are a visual display board that should be situated in a position near the main entrance that is clearly visible from the outside of the building. When the detection/alarm system is activated, the remote panel indicates the affected zone and whether the status is 'warning' or 'alarm'. This allows the emergency services or others to assess the situation without entering the building.

The users of a building that is equipped with a monitoring and alarm system should be made fully aware of its function, the procedures to be followed in case of alarms and the need for regular maintenance.

The present authors' experience with examples of these types of system (and active venting systems) that have been operational for a number of years indicates that a change of key staff in the organisation that looks after a building can have serious implications. It can result in the importance of the gas protection system not being recognised by the management team after a few years and maintenance may be neglected. Robust procedures to try and prevent this occurring are therefore required (e.g. regular inspections by the local authority or other body).

SUMMARY: Methods of gas protection

The most commonly used protection methods for developments can be divided into two groups:

- In ground vertical barriers
- Protection installed in buildings

Current UK construction practice adopts the concept of multiple gas protection measures, forming a gas control system, since no one protective measure is immune from factors unknown or outside the control of the designer. Typically, protection measures increase in number and robustness as the potential risk increases.

In ground barriers are used to prevent lateral gas migration from off-site sources and can be either impermeable or venting systems. These generally extend to an impermeable stratum or groundwater to cut off the migration pathway. There are various methods of forming such barriers but all vent systems should be designed to provide sufficient capacity to deal with the anticipated gas flows.

The main methods of protecting buildings from gas ingress are: the provision of a robust floor slab and gas-resistant barrier across the floor slab combined with an underfloor passive venting or positive pressurisation system. For commercial developments on more highly gassing sites additional protection may be provided by using active ventilation with monitoring and alarm systems as a backup to passive systems.

Reinforced concrete cast in situ suspended floor slabs can give reasonable gas resistance if they are detailed correctly. Block and beam slabs used in many housing developments have minimal resistance to ground gas ingress. The gas resistance of floor slabs is therefore improved by providing gas-resistant membranes. There are many different types of membrane

---

on the market but in practice because of the nature of construction sites the durability, survivability and robustness of membranes together with correct installation have the most impact on the effectiveness of the membrane in preventing gas entering a building. Inspection and quality assurance during installation are far more important than variations in methane permeability between different types of gas membrane.

Ventilation under the floors of buildings can be through open voids or via void formers below larger floor slabs. Systems should be designed to dilute gas concentrations to safe levels using natural ventilation wherever possible. Basement car parks can also provide this function. Positive pressurisation is a particular form of active system that works by pumping air into the void or void former below a floor slab to create a zone of pressure that is greater than the pressure of the gas in the ground. This prevents gas migrating to below the floor slab.

Gas monitoring and alarms are sometimes used to give added assurance that other methods of protection are working correctly. They can also be used to retrofit into existing buildings where it may be difficult to install other types of protection, or where emergency protection is required to be installed quickly before other protection systems can be installed. Because of the need for regular maintenance they are not appropriate in housing, except as an emergency measure in existing developments.

CHAPTER EIGHT

Design of protection measures

8.1 Defining the scope of gas protection

The first step in designing gas protection systems is to define the scope of the system required (i.e. what different types of protection will be used), based on the results of the risk assessment described in Chapter 6. Once the scope is defined detailed design of individual elements such as the venting or membranes can be completed. The latest CIRIA guidance has two methods of defining the scope of gas protection: one is for low-rise housing and a ventilated underfloor void (minimum 150 mm high); the other is for any other type of development (see Figure 8.1). Once the scope of the protection has been defined (i.e. the number of individual methods required) each element can be designed in detail (see Figure 8.2).

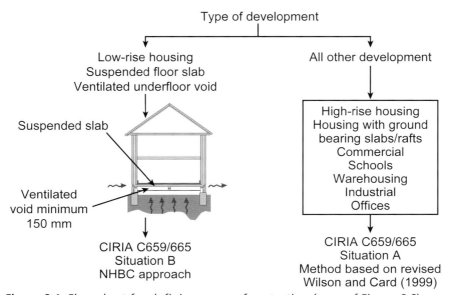

Figure 8.1 *Flow chart for defining scope of protection (copy of Figure 6.6)*

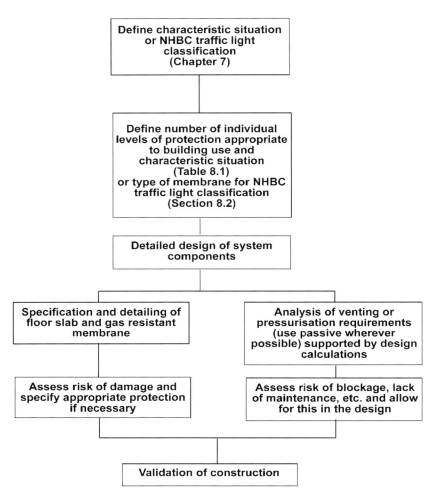

Figure 8.2 *Design process*

8.2 Low-rise housing with gardens (Situation B in CIRIA C659/665)

The system for low-rise housing was developed for the NHBC (Boyle and Witherington, 2005). The method is specific to low-rise housing that has a ventilated underfloor void that is at least 150 mm high. The GSV is used to determine the colour-coded classification of a site and the corresponding level of protection (Table 6.4). However, it is difficult to characterise housing stock into simple categories and care should be taken that the approach is applicable to a particular development. It has also been noted previously that the use of suspended floor slabs reduces the risk of damage to a gas-resistant membrane caused by differential settlement between a ground-bearing slab and the walls

Figure 5.2 *Examples of well spacing*

Figure 5.2 *Examples of well spacing (continued)*

Table 5.3 *Suggested periods and frequency of monitoring (based on Wilson and Haines, 2005)*

		Generation potential of source (see Table 3.1)				
		Very low	Low	Moderate	High	Very high
Sensitivity of development	Low	$\frac{4}{1}$	$\frac{6}{2}$	$\frac{6}{3}$	$\frac{12}{6}$	$\frac{12}{12}$
	Moderate	$\frac{6}{2}$	$\frac{6}{3}$	$\frac{9}{6}$	$\frac{12}{12}$	$\frac{24}{24}$
	High	$\frac{6}{3}$	$\frac{9}{6}$	$\frac{12}{6}$	$\frac{24}{12}$	$\frac{24}{24}$

Notes
1. First number is typical number of readings and second number is typical period in months, for example $\frac{4}{1}$ (four sets of readings over one month)
2. At least two sets of readings must be at low and falling atmospheric pressure (but not restricted to periods below <1000 mb) (known as worst case conditions (see RSK ENSR, 2005))
3. The frequency and period stated are considered to represent typical requirements. Depending on site specific circumstances, fewer or additional readings may be required (i.e. only such variations are subject to site specific justification
4. Historical data should be used as part of the data set
5. Not all sites will require gas monitoring. However, this would need to be justified with good evidence
6. The requirements apply only where a credible pollutant linkage to a source is identified
7. Placing a high sensitive end use on a high hazard site is not normally acceptable unless the source is removed or treated to reduce its gassing potential. Under such circumstances, long-term monitoring may not be appropriate or required
8. The generation potential of the gas source can be estimated from organic carbon content and compared to typical values given by the Environment Agency (2004a)
9. This table is taken from CIRIA C659 but with additional notes provided. The format of the table was changed in CIRIA C665 but the technical data remains the same

Table 6.4 *NHBC classification for low rise housing (150 mm void) (Boyle and Witherington, 2007; CIRIA, 2006)*

Traffic light classification	Methane[1]		Carbon dioxide[1]	
	Typical maximum concentration[5] (%v/v)	Gas screening value[2,4,6] (l/h)	Typical maximum concentration[5] (%v/v)	Gas screening value[2,3,4,5] (l/h)
Green				
	1	0.16	5	0.78
Amber 1				
	5	0.63	10	1.56
Amber 2				
	20	1.56	30	3.10[7]
Red				

1 The worst-case ground gas regime identified on the site, either methane or carbon dioxide, recorded in the worst temporal conditions will be the decider for which Traffic Light and GSV is allocated.

2 Generic GSVs are based on guidance contained within the latest revision of the Department of the Environment and the Welsh Office (2004 Edition) *The Building Regulations: Approved Document C* used a sub-floor void of 150 mm thickness.

3 The small room e.g. a downstairs toilet, with dimensions of 1.5 × 1.5 × 2.5 m, with a soil pipe passing through the sub-floor void.

4 The GSV (in litres per hour) is as defined in Wilson and Card (1999) as the borehole flow rate multiplied by the concentration of the particular gas being considered.

5 The "Typical Maximum Concentrations" can be exceeded in certain circumstances should the Conceptual Site Model indicate it is safe to do so. This is where professional judgement will be required, based on a thorough understanding of the gas regime identified at the site where monitoring in the worst temporal conditions has occurred.

6 The Gas Screening Value thresholds should not generally be exceeded without the completion of a detailed gas risk assessment taking into account site-specific conditions.

7 This value is taken from the March 2007 edition of Boyle and Witherington. It is quoted as 3.13 l/h in CIRIA C665.

8 The notes are taken from CIRIA C665 and are slightly different to those provided in Boyle and Witherington.

Table 6.5 *Characteristic situations (Wilson et al., 2006/2007, modified from Wilson and Card, 1999)*

Character-istic Situa-tion (CIRIA 149)[2+6]	Risk classifi-cation	Gas screen-ing value (CH_4 or CO_2) (l/h)[1+5]	Additional factors	Typical source of generation[3]
1	Very low risk	<0.07	Typically methane not to exceed 1 percent by volume and/or carbon dioxide not to exceed 5 percent by volume otherwise consider increase to situation 2	Natural soils with low organic content "Typical" made-up ground
2	Low risk	<0.7	Borehole air flow rate not to exceed 70 l/h otherwise consider increase to character-istic situation 3	Natural soil, high peat/organic content "Typical" made-up ground
3	Moder-ate risk	<3.5		Old landfill, inert waste, minework-ing flooded
4	Moderate to high risk	<15	Quantitative risk assessment required to evaluate scope of protective measures	Mineworking – susceptible to flooding, com-pleted landfill see Waste Manage-ment Paper 26 (D.E, 1994)
5	High risk	<70		Mineworking unflooded inactive with shallow work-ings near surface
6	Very high risk	>70		Recent landfill site

1 Gas screening value (litres of gas/hour) is calculated by multiplying the maximum gas concentration (%) by the maximum mea-sured borehole flow rate
2 Site characterisation should be based on gas monitoring of concentrations and bore-hole flow rates for the minimum periods defined in Section 5
3 Source of gas and generation potential/performance should be identified
4 Ground gas investigation to be in accor-dance with current good practice
5 If there is no detectable flow use he limit of detection of the instrument
6 In this table, the column which compares characteristic situations with the DETR classification has been omitted. This is the version from CIRIA C665 and the notes are slightly different to the version in C659

Box 6.4 Comparison of screening methods for housing (up to Characteristic Situation 4)

CIRIA Situation A and BS 8485 (methane and carbon dioxide)

Gas screening value l/h

Characteristic situation	Lower boundary	Upper boundary	Other criteria
1	0	0.07	< 1% methane < 5% carbon dioxide
2	0.7	0.7	Flow rate < 70l/h
3	3.5	3.5	
4	15	15	Quantitative risk assessment required

NHBC Traffic lights (methane)

Gas screening value l/h

Traffic light	Lower boundary	Upper boundary	Other criteria
Green	0	0.16	< 1% methane
Amber 1	0.16	0.63	< 5% methane
Amber 2	0.63	1.56	< 20% methane
Red	1.56		

NHBC Traffic lights (carbon dioxide)

Gas screening value l/h

Traffic light	Lower boundary	Upper boundary	Other criteria
Green	0	0.78	< 5% carbon dioxide
Amber 1	0.78	1.56	< 10% carbon dioxide
Amber 2	1.56	3.13	< 30% carbon dioxide
Red			

Protection Requirements

Housing

CIRIA Situation A (assuming applied to low rise housing instead of NHBC)	NHBC traffic light (methane)
No protection (0)	Ventilated underfloor void, block and beam
Level 2 Venting cast insitu slab and 1200g mem or Venting, block and beam and 2000g mem Assumes very good venting	Venting, block and beam and 1200g mem Assumes very good venting
Level 2 Venting, any slab, proprietary gas mem Assumes very good venting	Void and gas resistant membrane (it is implied that the membrane is installed by specialist contractors and that it is tested and independently inspected) Assumes very good venting
CS4 AND ABOVE NOTE THAT COMPREHENSIVE RISK ASSESMENTIS REQUIRED TO CONFIRM MEASURES	NOT SUITABLE WITHOUT RISK ASSESSMENT

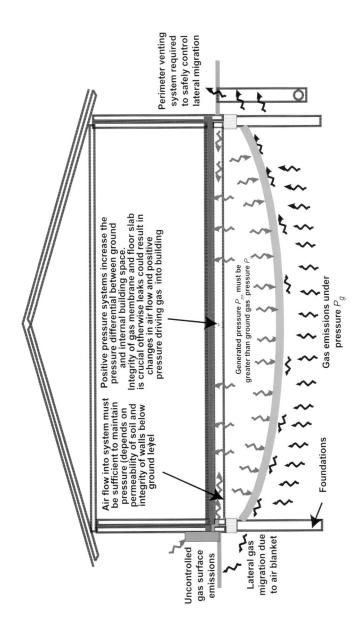

Perimeter venting system required to safely control lateral migration

Positive pressure systems increase the pressure differential between ground and internal building space.
Integrity of gas membrane and floor slab is crucial otherwise leaks could result in changes in air flow and positive pressure driving gas into building

Air flow into system must be sufficient to maintain pressure (depends on permeability of soil and integrity of walls below ground level

Generated pressure P_m must be greater than ground gas pressure P_g

Gas emissions under pressure P_g

Foundations

Uncontrolled gas surface emissions

Lateral gas migration due to air blanket

Figure 7.13 *Positive pressurisation*

Table 8.1 *Scope of protection measures (CIRIA, 2006)*

Characteristic Situation	Residential building (Not low rise traditional housing)[1]		Office/commercial/industrial development	
(From Table 6.5)	Number of levels of protection	Typical scope of protective measures	Number of levels of protection	Typical scope of protective measures
1	None	No special precautions	None	No special precautions
2	2	a) reinforced concrete cast *in situ* floor slab (suspended, non suspended or raft) with minimum 1200g DPM and underfloor venting b) block and beam or pre-cast concrete slab and 2000g DPM/or reinforced gas membrane and underfloor venting All joints and penetrations sealed	1 to 2	a) reinforced concrete cast *in situ* floor slab (suspended, non suspended or raft) with minimum 1200g DPM b) block and beam or precast concrete slab and minimum 2000g DPM/or reinforced gas membrane c) possibly under-floor venting or pressurisation in combination with a) and b) depending on use All joints and penetrations sealed
3	2	All types of floor slab as above. Proprietary gas resistant membrane and passively ventilated or positively pressurised underfloor sub-space. All joints and penetrations sealed	1 to 2	All types of floor slab as above. Proprietary gas resistant membrane and passively ventilated underfloor sub-space or positively pressurised underfloor sub-space. All joints and penetrations sealed

Table 8.1 *Scope of protection measures (CIRIA, 2006) (continued)*

Characteristic Situation	Residential building (Not low rise traditional housing)[1]		Office/commercial/industrial development	
4	3	All types of floor slab as above. Proprietary gas resistant membrane and passively ventilated underfloor subspace or positively pressurised underfloor sub-space, oversite capping or blinding and in ground venting layer. All joints and penetrations sealed	2 to 3	All types of floor slab as above. Proprietary gas resistant membrane and passively ventilated or positively pressurised underfloor sub-space with monitoring facility. All joints and penetrations sealed
5	4	Reinforced concrete cast *in situ* floor slab (suspended, non suspended or raft). Proprietary gas resistant membrane and ventilated or positively pressurised underfloor sub-space, oversite capping and in ground venting layer and in ground venting wells or barriers. All joints and penetrations sealed	3 to 4	Reinforced concrete cast *in situ* ground slab (suspended, non suspended or raft). Proprietary gas resistant membrane and passively ventilated or positively pressurised underfloor sub-space with monitoring facility. In ground venting wells or barriers. All joints and penetrations sealed

Table 8.1 *Scope of protection measures (CIRIA, 2006) (continued)*

Characteristic Situation	Residential building (Not low rise traditional housing)[1]		Office/commercial/industrial development	
6	5	Not suitable unless gas regime is reduced first and quantitative risk assessment carried out to assess design of protection measures in conjunction with foundation design.	4 to 5	Reinforced concrete cast *in situ* ground slab (suspended, non suspended or raft). Proprietary gas resistant membrane and actively ventilated or pressurised underfloor sub-space, with monitoring facility. In ground venting wells and reduction of gas regime.

All joints and penetrations sealed |

1 For low-rise traditional housing see Table 6.4 and Section 8.2.
2 Typical scope of protective measures may be rationalised for specific developments on the basis of quantitative risk assessments.
3 Note the type of protection is given for illustration purposes only. Individual site specific designs should provide the same number of separate protective methods for any given characteristic situation.
4 In all cases there should be minimum penetration of ground slabs by services and minimum number of confined spaces such as cupboards above the ground slab. Any confined spaces should be ventilated.
5 Foundation design must minimise differential settlement particularly between structural elements and ground-bearing slabs.
6 Commercial buildings with basement car parks, provided with ventilation in accordance with the Building Regulations, may not require gas protection for characteristic situations 3 and 4. However, features such as stair wells, lift pits, bin stores, etc that are located at basement floor level will require separate consideration and may still require some form of protection.
7 Floor slabs should provide an acceptable formation on which to lay the gas membrane. If a block beam floor is used it should be well detailed so it has no voids in it that membranes have to span and all holes for service penetrations should be filled. The minimum density of the blocks should be 600 kg/m^3 and the top surface should have a 4:1 sand cement grout brushed into all joints before placing any membrane (this is also good practice to stabilise the floor and should be carried out regardless of the need for gas membranes).
8 The gas resistant membrane can also act as the damp proof membrane.

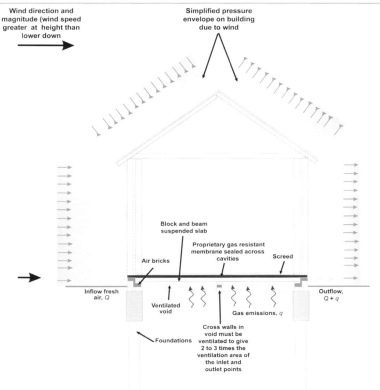

Figure 8.3 *Passive venting of an underfloor void*

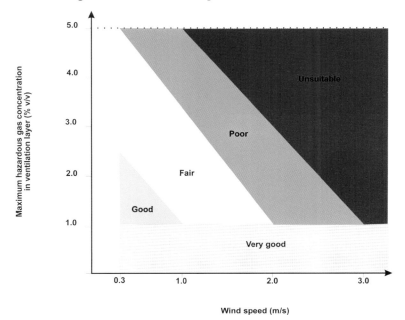

Figure 8.4 *Performance assessment criteria of ventilation layer for methane hazard (DETR, 1997)*

a

b

Figure 9.1 *Construction problems*

Figure 9.2 *Membrane torn and wter sevice fixed by nailing through mebrane*

Yes–there is membrane under there!

Figure 9.3 *Scaffolding and general dirt on top of exposed membrane*

Figure 9.4 *Membrane installation (note some sealing is yet to be completed) Picture supplied by Sarah Stroud, RSK*

Figure 9.5 *Poor membrane installation*

of a building and so they are the preferred slab solution on sites affected by ground gas, unless it can be demonstrated that movement of the slab will not damage the membrane.

Once the traffic light classification has been determined the appropriate level of protection can be defined using the NHBC guidance or CIRIA C659/665 (Boyle and Witherington, 2007; Wilson *et al.*, 2006, 2007). The requirements are summarised below:

Green: No protection required (over and above the underfloor ventilated void that is assumed in the use of this approach).

Amber 1: Low-level gas protection using a membrane and ventilated underfloor void in accordance with BRE Report 414, ventilated void to give at least one air change per day.

Amber 2: High-level gas protection using a membrane and ventilated underfloor void in accordance with BRE Report 414, ventilated void to give at least one air change per day. Membrane to be installed by specialist contractor and the membrane and ventilation to be certified as being installed correctly. This requires that the membrane is installed under a CQA regime by a specialist and is tested after installation (Appendix E3 of Boyle and Witherington, 2007). In effect this means that specialist gas-resistant membranes will be required.

Red: Standard low-rise residential housing with gardens is not acceptable without further detailed assessment of the ground gas regime and the risks it poses and/or mitigation measures to reduce and/or remove the source of gas.

The authors of this handbook also recommend that the whole installation of membrane and passive venting is independently inspected at suitable points during construction (for both Amber 1 and 2).

Box 8.1 Using the NHBC method (Situation B in CIRIA C659/665)

A site has given the following maximum results from gas monitoring (These results have been assessed against the desk study information and are considered to be representative of the potential sources on or adjacent to the site.):

Maximum methane concentration = 0.6% v/v
Maximum carbon dioxide concentration = 6.8% v/v
Maximum borehole flow rate = 2 l/h

So the GSVs are:
For methane GSV = $0.006 \times 2 = 0.012$ l/h. From Table 6.4 = green as GSV is less than 0.13 l/h and concentration of methane is less than 1% For carbon dioxide = $0.068 \times 2 = 0.14$ l/h = amber 1 because although GSV is less than 0.78 l/h the concentration of carbon dioxide is higher than 5% and this increases the classification from green to amber 1.

Therefore the development should have the following protection measures to the houses:

Low-level gas protection using a membrane and ventilated underfloor void in accordance with BRE Report 414, ventilated void to give at least one air change per day.

CIRIA Report C659 provides more examples.

8.3 All other types of development (Situation A in CIRIA C659/665)

For all other types of development the scope of gas protection measures is determined by using a modified version of the approach described by Wilson and Card (1999). Once the characteristic situation for the site is known then the scope of protection measures can be determined using Table 8.1 (see colour section). It should be noted that this table has been slightly modified, with additional notes and clarification regarding membranes.

The levels of redundancy referred to in Table 8.1 means providing a number of protective elements that are each capable of protecting a building on their own. In case of failure or damage of one element the remaining element(s) continue to protect the building.

When using Table 8.1 the following approach has been adopted in practice:

- The membrane and floor slab together are assumed to provide one level of protection together (i.e. they provide a barrier, also see note below on membranes)
- Ventilation design below housing can usually achieve very good performance (see later in this section). For large commercial and industrial buildings ventilation is usually designed to give good performance in accordance with the Partners in Technology report. Commercial and industrial buildings also widely use void formers to make the underfloor void resulting in a lower void depth than for housing. Thus the time to fill to 5% in no wind conditions is usually lower
- For industrial buildings in characteristic situation 2, where underfloor venting is not provided, a nominal gas release system may be required for larger buildings to prevent the build up of gas pressure below the floor slab (e.g. by using pipes or strips of geocomposite in gravel)

8.3.1 Membranes and the use of Table 8.1

The membrane requirements in Table 8.1 are only indicative and may be changed subject to clear and rational justification. The purpose of Table 8.1 is to show common solutions for each characteristic situation. The solutions are based on the assumption that achieving a good membrane installation is more important with leaky floor slabs such as block and beam. Thus block and beam slabs require a more robust membrane than a cast in situ reinforced concrete floor slab (see Chapter 7 for more discussion on membranes). Regardless of material type all membranes should be inspected after installation and protected from damage (see Chapter 9). If there is likely to be any differential movement between the floor slab and the wall foundations this must be allowed for in the design and specification of the gas membrane.

The membrane requirements in Table 8.1 are consistent with the advice in Chapter 7. The NHBC guidance requires membranes that are installed on Amber 2 sites to be integrity tested. It is recommended that this requirement is also applied to membranes on sites that fall in situation 3 and above in the modified Wilson and Card approach (Situation A in CIRIA, 2006).

For low-risk sites if 1200 g DPM is used below a substantial floor slab it needs very careful installation and possibly protection from follow on trades using protective boards or fleeces. It will also require very close independent inspection. Thus it can often be cost effective to use a more robust membrane with less protection.

Box 8.2 Use of modified Wilson and Card method (CIRIA, 2006)

A site is proposed for a shopping centre development and has a maximum gas concentration of 26% and borehole flow rate of 0.1 l/h. The source of the gas is a thick layer of alluvium with peat lenses and groundwater levels in the area are known to rise and fall quickly in response to rainfall events. The development will have a suspended cast in situ reinforced concrete floor slab that is 200 mm thick.

Eight sets of gas monitoring results are available but these have been taken when there been little rainfall and groundwater levels have remained constant.

So, we can consider that the borehole flow rates from the gas monitoring (0.1 l/h) are not representative. We can estimate that if groundwater levels rise we may obtain higher rates in the short term (say, for up to 8 h). So, we will estimate that 2 l/h borehole flow rate is more representative of the site.

Using this data the GSV = 2 l/h × 26% = 2 × 0.26 = 0.52 l/h.

From Table 6.5, we can see that the characteristic situation is 2.

Therefore the gas protection for the development will require one or two levels of redundancy. As the shopping centre is a moderate sensitivity end use with lots of people using it and the potential for small spaces at ground level we will provide two levels of protection.

In this case the 200 mm thick reinforced concrete cast in situ slab gives a good inherent resistance against gas ingress. This can be enhanced with a 1200 g DPM and a passive underfloor venting layer in accordance with Table 8.1. However, it should be noted that 1200 g DPM will require sealing with tape at all joints and to service penetrations and across any cavities (if present). It will also require independent inspection to confirm that it has been installed correctly and without defects.

8.4 British Standard BS 8485

The approach from CIRIA Report C659 can be extended to look at a wider range of uses and to provide a more flexible choice of protection measures, for a given characteristic situation (again looking at methane and carbon dioxide only). This is the approach adopted in British Standard BS 8485 and Table 8.2 provides a minimum score for a range of end uses. The score indicates the comparative level of risk for a site based on the use and the characteristic gas situation. This is simply based on the assumption that as the characteristic situation increases the risk increases and that as the sensitivity of a building decreases the risk decreases.

Table 8.2 *Comparative risk score based on design gas regime and sensitivity of end use (refer to BS 8485 for the most up-to-date version)*

Characteristic gas situation CS	NHBC Traffic light	Required gas protection			
		Non-managed property e.g. private housing	Public buildings[A]	Commercial buildings	Industrial buildings[E]
1	Green	0	0	0	0
2	Amber 1	3	3	2	1[C]
3	Amber 2	4	3	2	2
4	Red	6[D]	5[D]	4	3

Table 8.2 *Comparative risk score based on design gas regime and sensitivity of end use (refer to BS 8485 for the most up-to-date version) (continued)*

Characteristic gas situation CS	NHBC Traffic light	Required gas protection			
		Non-managed property e.g. private housing	Public buildings[A]	Commercial buildings	Industrial buildings[E]
5		Generally not suitable without changing the characteristic gas situation, e.g. pathway intervention or in ground venting.	6[E]	5	4
6		Generally not suitable without changing the characteristic gas situation, e.g. pathway intervention or in ground venting.	Generally not suitable without changing the characteristic gas situation, e.g. pathway intervention or in ground venting.	7	6

A) Public buildings include, for example, managed apartments, schools and hospitals.

B) Industrial buildings are generally open and well ventilated. However, areas such as office pods may require a separate assessment and may be classified as commercial buildings, requiring a different scope of gas protection to the main building.

C) Maximum methane concentration 20 percent otherwise increase protection to CS3 level.

D) Residential building on higher Traffic Light/CS sites is not recommended unless the type of construction or site circumstances allow additional levels of protection to be incorporated, e.g. high-performance ventilation or pathway intervention measures, and an associated sustainable system of management of maintenance of the gas control system, e.g. in institutional and/or fully serviced contractual situations.

E) Consideration of issues such as the ease of evacuation and how false alarms will be handled are required when completing the design specification of any protection scheme.

As usual with ground gas there will be judgement required when using Table 8.1. The designer must be sure that the chosen end use classification is appropriate to the particular site design. For example the industrial buildings section assumes that there are no small confined spaces. In most sites industrial or warehouses have office pods or toilets which may need to be classified differently and provided with different protection measures to the main warehouse space.

Even though this is included in a British Standard and it was developed to give similar results to the CIRIA C665 and NHBC approaches, it is a new and unproven rating system and extreme care must be taken with its use until it becomes well established. The use of the risk scores and component rankings must be fully justified by consultants using this method.

The traffic light classification is provided to give an indicative comparison only as the actual boundaries between the levels are different (see Box 6.3). The traffic light classification is restricted to low-rise housing with a ventilated underfloor void.

The range of protection measures that are available are given a ranking score in Table 8.3 and the combined score for the protection measures that are provided must meet or exceed the required risk score in Table 8.2. It should be noted that there is a hierarchy of preference for the different techniques that must be adhered to. The British Standard requires the protection measures to be used in the following order of preference:

(1) Standard protection measures
   - Concrete floor slab
   - Underfloor natural (passive) venting wherever possible (or car park). Where passive ventilation is not feasible or cost effective use active venting or positive pressurisation
   - Barrier membranes
   - Oversite capping or concrete blinding below underfloor voids
(2) Reduction in gas regime using in ground barriers to prevent migration from an off-site source or in ground venting to reduce the risk of surface emissions from an on-site source
(3) Enhanced protection (for higher gas risks: characteristic situation 5 and 6)
   - Active underfloor ventilation
   - Monitoring and alarm systems

Table 8.3 *Ranking scores for protective measures from BS 8485: 2007*

Protection element/system		Score	Comments
a) Venting/dilution (see Annex A)			
Passive subfloor ventilation (venting layer can be a clear void or formed using gravel, geocomposites, polystyrene void formers, etc.)^A	Very good performance	2.5	Ventilation performance in accordance with Annex A of the British Standard.
	Good performance	1	If passive ventilation is poor this is generally unacceptable and some form of active system will be required

Table 8.3 *Ranking scores for protective measures from BS 8485: 2007 (continued)*

Protection element/system	Score	Comments
Subfloor ventilation with active abstraction/pressurisation (venting layer can be a clear void or formed using gravel, geocomposites, polystyrene void formers etc.)[A]	2.5	There have to be robust management systems in place to ensure the continued maintenance of any ventilation system Active ventilation can always be designed to meet good performance Mechanically assisted systems come in two main forms: extraction and positive pressurisation
Ventilated car park (basement or undercroft)	4	Assumes car park is vented to deal with car exhaust fumes, designed to Building Regulations Document F and IStructE guidance.
b) Barriers		
Floor slabs		
Block and beam floor slab	0	It is good practice to install ventilation in all foundation systems to effect pressure relief as a minimum. Breaches in floor slabs such as joints have to be effectively sealed against gas ingress in order to maintain these performances.
Reinforced concrete ground bearing floor slab	0.5	
Reinforced concrete ground bearing foundation raft with limited service penetrations that are cast into slab	1.5	
Reinforced concrete cast *in situ* suspended slab with minimal service penetrations and water bars around all slab penetrations and at joints	1.5	
Fully tanked basement	2	
c) Membranes		
Taped and sealed membrane to reasonable levels of workmanship/in line with current good practice with validation[B), C)]	0.5	The performance of membranes is heavily dependent on the quality and design of the installation, resistance to damage after installation, and the integrity of joints
Proprietary gas-resistant membrane to reasonable levels of workmanship/in line with current good practice under independent inspection (CQA)[B), C)]	1	
Proprietary gas resistant membrane installed to reasonable levels of workmanship/in line with current good practice under CQA with integrity testing and independent validation	2	

Table 8.3 *Ranking scores for protective measures from BS 8485: 2007 (continued)*

Protection element/system		Score	Comments
d) Monitoring and detection (not applicable to non-managed property, or in isolation)			
Intermittent monitoring using hand-held equipment		0.5	
Permanent monitoring and alarm system[A]	Installed in the underfloor venting/ dilution system	2	Where fitted, permanent monitoring systems ought to be installed in the underfloor venting/ dilution system in the first instance but can also be provided within the occupied space as a fail safe
	Installed in the building	1	
e) Pathway intervention			
Pathway intervention		–	This can consist of site protection measures for off-site or on-site sources (see Annex A of the Standard)

[A] It is possible to test ventilation systems by installing monitoring probes for post-installation validation

[B] If a 1 200 g DPM material is to function as a gas barrier it should be installed according to BRE 212/BRE 414, being taped and sealed to all penetrations.

[C] Polymeric materials > 1 200 g can be used to improve confidence in the barrier. Remember that their gas resistance is little more than the standard 1 200 g (proportional to thickness) but their physical properties mean that they are more robust and resistant to site damage.

NOTE: In practice the choice of materials might well rely on factors such as construction method and the risk of damage after installation. It is important to ensure that the chosen combination gives an appropriate level of protection.

The scores in each category are only comparative for the individual elements within that category, with a higher score only indicating a more robust solution. An increase in score from 1 to 2 does not mean that the component with a score of 2 is twice as effective.

In some cases it is possible to provide protection using a robust reinforced concrete raft foundation and a gas-resistant membrane. In these cases an underfloor gas pressure relief layer will be required to ensure that gas pressures cannot build up. Such a system is designed to prevent gas pressure build up rather than to dilute gas to a specified equilibrium concentration. When designing such a solution care must be taken that the emissions at the outlets will not cause an unacceptable risk. An example showing the use of the extended method is given in Box 8.3.

Box 8.3 Use of BS 8485 method

A site is proposed for a shopping centre development and has a maximum gas concentration of 26% and borehole flow rate of 0.1 l/h. The source of the gas is a thick layer of alluvium with peat lenses and groundwater levels in the area are known to rise and fall quickly in response to rainfall events.

Eight sets of gas monitoring results are available but these have been taken when there has been little rainfall and groundwater levels have remained constant.

So we can consider that the borehole flow rates from the gas monitoring (0.1 l/h) are not representative. We can estimate that if groundwater levels rise we may obtain higher rates in the short term (say, for up to 8 h). So we will estimate that 2 l/h borehole flow rate is more representative of the site.

Using this data the GSV = 2 l/h × 26% = 2 × 0.26 = 0.52 l/h.

From Table 6.5 the characteristic situation is 2.

Therefore from Table 8.2 the buildings require a score of 2 as the shopping centre is a moderate sensitivity end use with lots of people using it and the potential for small spaces at ground level.

From Table 8.3 we can utilise the following information:

Item	Score
A reinforced concrete cast in situ ground bearing slab	0.5
A 1200 g DPM that is tape sealed and installed by main contractor (but with independent inspection and protection from damage)	0.5
A passive underfloor venting layer designed for good performance (due to width of building and moderate sensitivity)	1
Total score	2

Comparison with CIRIA Situation A approach

CIRIA Situation A: Development is classified as 'office/commercial/industrial development'

For characteristic situation 2 the development would require a reinforced concrete cast in situ floor slab (suspended, non-suspended or raft) with minimum 1200 g DPM. Underfloor venting, or pressurisation in combination with the slab and membrane, may possibly be required depending on use. In this case because of the intended end use the venting would be required (i.e. the same scope of protection is required from both methods).

When using the extended system the final result must be assessed to ensure it does not result in inappropriate combinations of protection measures (e.g. using monitoring and alarms as the only method of protection in a new building).

The extended system should be comparable to the other methods described in CIRIA Report C659.

8.5 Key points for design of gas protection systems

- A combination of measures is usually used to minimise the risks associated with ground gas. This is so that if one element of the protection fails the others will continue to protect the development. Only where the risk posed by the presence of ground gas is very low will a single level of protection be suitable
- The scope of measures for any site should be related to the risk assessment, with an increasing element of redundancy (or extra protection methods) provided as the risk increases
- Use of precautionary gas protection without a site investigation is a valid approach for low-risk sites, provided that a desk study is undertaken to show there are no significant gas sources that could affect the site
- If the gas source is adjacent to the development site the design philosophy should be based primarily on keeping gas out of the site using in ground barriers, and then protecting the buildings. If that is not possible (if the source is below the site) then the main gas protection should be below the building. Passive ventilation to give a clean zone of air around a building is the preferred primary method of protection wherever possible, with secondary protection provided by a gas membrane
- The requirement for redundancy in gas protection means that generally at least two different methods are required to ensure protection is maintained if one method fails (e.g. a membrane and ventilated void)
- No single method can provide two types of protection are untrue. For example a positive pressurisation system does not provide two levels of protection (a barrier and a ventilation system). If it stops working a second level of protection, such as a membrane, is required to protect a building

- In most cases of protection focused on the building, passive venting and dilution should be the primary protection method. Gas-resistant membranes should be provided as secondary protection to allow for the fact that ventilation may stop working. Passive venting is preferred as it is generally robust and can be designed to continue working when a number of vents are blocked. Any active system (venting or pressurisation) is much more reliant on continued maintenance. Active systems should be designed to act as passive systems if the fans stop working
- For larger buildings the use of passive venting becomes more difficult and active venting or pressurisation may be required
- Internal monitoring should be considered as a last resort, not a means of protection. The best place for monitoring points is in the underfloor void or vent system
- Where service ducts are sealed to a gas membrane the internal space in the duct must be sealed with expanding foam and a mastic sealant collar. Omission of this detail has been observed to be the cause of failure of gas protection systems on several sites
- There is often concern about the durability of membranes at sites where hydrocarbon contamination is present. The membranes are normally located in clean material above the soil contamination and thus are not in direct contact with it and only come into contact with vapours. For most sites the site-specific clean up concentrations of hydrocarbons will be such that any vapour coming into contact with a membrane will be at extremely low concentrations. These very low concentrations should not adversely affect the materials used in gas- and vapour-resistant membranes, such as PE and PP. If there is any doubt a site-specific risk assessment should be completed

8.6 Detailed analysis of venting and positive pressurisation systems

8.6.1 Passive venting of voids

Passive systems should be designed to dilute gases to a safe concentration at the outlet (typically 0.25–1% for methane and carbon dioxide, but this will be gas specific. Similar values can be derived for other gases and vapours using the data in Tables 3.2 and 3.3). Simply venting gases from below a slab is not usually acceptable, except for very low-risk sources of gas, because the gas could reach hazardous concentrations at the vents.

Passive ventilation relies on wind-induced pressure and suction on the sides of buildings that drives fresh air flow through the underfloor void and dilutes any gas that is emitted from the ground. The actual magnitude of the pressure and whether or not suction occurs on a particular elevation depends on the height and orientation of the building in relation to the wind direction and the location of the inlet and outlet points (see Figure 8.2). The wind direction and speed will vary (BS 5925: 1991 provides guidance) and this should be allowed for in the design:

- For simple airbrick vents or similar there is usually some suction or pressure on the wall, regardless of wind direction. This is allowed for by using a very low value of ΔP, typically 0.4–0.6 (see Figure 8.2 and Box 8.5)
- For vent stacks a similar approach is adopted using a low value of ΔP. Rotating cowls can also be used to enhance air flow up the stacks. The cowls operate regardless of wind direction

In both cases the effect of differential temperatures between the outside air and the underfloor void are ignored. In practice this can induce significant air flow.

Suction effects are greatest at the edges of roofs, so high-level vent stacks are usually taken to just above eaves level in order to maximise the suction up the stacks. Further information is provided in CIRIA Report 149.

Passive venting is a proven technology and it has been demonstrated to be effective in practice, both by computer modelling and by measuring the gas concentrations and flow rates in constructed systems (Pecksen, 1981; Wilson and Card, 1999; DETR, 1997).

Although it relies on a naturally variable driving force it is designed on con-servative principles (outlined previously) to ensure that it continues to work under very low wind speeds. The periods when it is not working will be short and within acceptable limits so that gas does not build up to unacceptable con-centrations. The maximum period of no wind assumed for design purposes is 10 h in the UK.

Long-term monitoring of gas concentrations and air flow rates through vents in geocomposite systems has demonstrated that the systems can be designed to work up to lengths of about 50 m on sites with low levels of gas. Gravel-based venting layers are much less effective. The design calculations have also been shown to give a reasonable indication of the likely performance of the venting layers.

To give added confidence in the performance of passive systems, gas monitor-ing can be undertaken in the underfloor voids on completion of construction (usually in the period between completion of the floor slab and handover of the building). Purpose-built monitoring points should be provided for this. On most sites the ground gas emissions vary and therefore to demonstrate that low gas concentrations in the void occur because the ventilation is working (rather than because there are no ground gas emissions) it is usually necessary to monitor the in ground gas concentrations and flow rates at the same time. If necessary, and appropriate, the system can be designed so that it can be improved by retrofitting active ventilation or additional ventilation points if the monitoring shows the passive venting is not working sufficiently well (see Section 8.6.6).

When designing or analysing any vent system (active as well as passive) the limiting factor that will determine the air flow through the whole system should be identified (e.g. the geocomposite layer, etc.). Careful detailing of inlet and outlet positions is required to promote smooth air flow and reduce the risk of dead spots or short-circuiting occurring.

For all types of underfloor venting system the venting area provided at the perimeter should be continued through any downstand beams or sleeper walls and should be much greater than the area of the side ventilation provided (typically two or three times greater).

8.6.2 Estimating passive ventilation performance

Passive ventilation performance can be estimated using simple calculations. The most commonly adopted approach is described and examples are given for open voids and air flow through a void former.

The fresh air flow, Q, required to dilute the gas emissions, q, in the void to a defined level can be estimated using the following equation from CIRIA Report 149. Fresh air flow required (total under whole building), Q is given by:

$$Q = q\{(100 - C_e)/C_e\}$$

Where q = surface emission rate of gas from the ground (total under whole building) (see Chapter 4) and C_e = equilibrium gas concentration in the void (in this case expressed as the % value not the mathematical value i.e. for 1% use 1 in the equation rather than 0.01) (see Box 8.4).

Box 8.4 Calculation of required fresh air flow rate

Surface emission rate of methane = 0.023 l/h/m²

Convert this to m³ = $\frac{0.023}{1000}$ = 2.3×10^{-5} m³/h/m²

Building is 20 m wide × 40 m long

So total surface emission rate of gas into void, q = 20 m × 40 m × 2.3 × 10^{-5} m³/h/m² = 0.0184 m³/h

Required design concentration is 1% v/v of methane in the void and at outlets

Design fresh air flow rate required, $Q = 0.0184 \text{ m}^3/\text{h} \times [(100 - 1)/1] = 0.0184 \times 99 = 1.8 \text{ m}^3/\text{h}$

The design of the ventilation system must provide this volume of air flow to provide sufficient dilution of the gas emissions. This applies to open void or void former systems.

8.6.3 Open void

The required fresh air flow rate determined using the equation from CIRIA 149 can be substituted in the equations from BS 5925: 1991 to determine the necessary ventilation area. Alternatively a pumping system can be sized to give this air flow. Box 8.5 gives an example of the use of the equations from BS 5925.

Box 8.5 Vent sizing

Determine the ventilation required for a housing development up to three storeys high located on the coast in Southampton, UK. The height of the vents is 0.15 m and the height of the building is 7 m. The required flow of fresh air through the void is 1.8 m³/h for a building that is 40 m long × 20 m wide. The underfloor void is 150 mm high and the surface emission rate is $2.3 \times 10^{-5} \text{m}^3/\text{h}/\text{m}^2$

From Figure 5 in BS 5925: 1991 Code of practice for ventilation principles and designing for natural ventilation, the hourly mean wind speed, $U_{50} = 4.5 \text{ m/s}$ (measured at 10 m height in open terrain).

Determine correction ratio from Table 9 in BS 5925: 1991.
Allow for the design wind speed being exceeded 80% of the time (i.e. this is the worst case value and gives the highest confidence that the passive system will operate) and consider an exposed coastal location
So factor = 0.56

$U_m = U_{50} \times 0.56 = 4.5 \times 0.56 = 2.52 \text{ m/s}$

Determine factors K and a from Table 8 in BS 5925: 1991 to allow for height of vent and nature of surrounding terrain.
Assume an urban environment so $K = 0.35$ and $a = 0.25$. These factors amend the mean hourly wind speed to allow for differing terrain and different heights. The pressure on the side of the building is governed by the height of the building but to be conservative use the height of the vent as the design height.

Therefore, from BS 5925: 1991 the reference wind speed is:

$u_r = u_m \times K \times z^a$ (where z = height of vent)

$\quad = 2.52 \times 0.35 \times 0.15^{0.25} = 0.55$ m/s

Now calculate the required vent A_w area to give flow of fresh air, Q. Assume that the discharge coefficient for a narrow opening, $C_d = 0.61$, which is a typical value for narrow openings from BS 5925: 1991. (This is a factor that correlates theoretical performance to actual performance.)

The orientation of buildings is not known so use a pessimistic value for ΔC_p, thus $\Delta C_p = 0.4$.

Area of ventilation required for whole building, A_w, is calculated using the following equation from BS 5925: 1991

$$A_w = \frac{Q}{U_r \times C_d \times \sqrt{\Delta C_p}} \times 10^6$$

$Q = 1.8$ m³/h and needs to be converted to m³/s $= \dfrac{1.8}{3600} = 0.0005$ m³/s

$$= \frac{0.0005}{0.55 \times 0.61 \times \sqrt{0.4}} \times 10^6 = 2360 \text{ mm}^2 \text{ (note the factor } 10^6 \text{ is used to con-}$$

vert m² to mm²)

See note below regarding allowance for vents in series.

This is equal to $\dfrac{2360 \text{ mm}^2}{40 \text{ m}} = \dfrac{60 \text{ mm}^2}{\text{m}}$. This is less than the minimum

venting area required in the Building Regulations of 1500 mm²/m so therefore the minimum vent area should be provided. This can be achieved by using normal air bricks with a vent area of 6000 mm² at 4 m centres.

Check the time taken to fill the void to 5% methane if there is no wind

Time to fill to 5% methane = (volume of void × 5%)/surface emission rate of gas below building

The plan area is 40 m × 20 m and the volume of the void is 40 × 20 × 0.15 = 120 m³

Time to fill = (120 m³ × 0.05)/(2.3 × 10⁻⁵ m³/h/m² × 40 m × 20 m) = 326 h

This is greater than the maximum period of still wind of 10 h reported in the Partners in Technology report and so is acceptable.

It has not been normal practice in ventilated void design to increase the vent area to allow for reduced effective vent area for vents in series (multiple pairs of vents located on opposing walls), as described in BS 5925: 1991, which would increase Aw by a factor of 1.4 in most cases. This is because the flow of air through a relatively small underfloor void with vents located at relatively frequent regular intervals is similar to airflow through a network of parallel pipes. The total air flow through the underfloor void is the sum of the flow through the individual pair of vents (on opposing sides of the building), but the overall pressure and friction loss is the same as that through any one pair of vents. Thus there is little or no reduction in effective vent area as a result of vents in series. Furthermore the calculations are widely accepted as already being conservative (e.g. ignoring temperature effects, using low values of ΔCp and using onerous values of gas emission rates). In the vast majority of cases the overriding requirement is to provide the minimum ventilation requirement of 1500 mm^2/m of wall (see Box 8.5). However, this needs to be borne in mind when trying to use less conservative estimates of gas emission rates or on higher risk sites where the required ventilation is approaching or higher than the minimum requirements.

8.6.4 Void formers

For passive venting of void formers the approach adopted requires two calculations. One to ensure there is sufficient air flow through the geocomposite and the second to size the vents, in the same way as for a normal ventilated void (see Box 8.6).

Box 8.6 Calculation of air flow through a geocomposite or polystyrene void former

Estimate the air flow through a 25 mm thick geocomposite layer that has a 10 m long flow path. There are various methods available to do this. One method is using Darcy's law (see Chapter 4). An example is given below.

Assume pressure drop across geocomposite due to wind is 5.4 Pa (DETR, 1997)

K_i = intrinsic permeability of geocomposite in m^2 = $1.2 \times 10^{-5}\ m^2$

γ = unit weight of air in N/m^3 = 13

μ = viscosity of air in Ns/m^2 = 1.75×10^{-5}

A = surface area over which flow occurs in m^2 = 0.025 × 1 m
  = 0.025 m^2

i = pressure gradient across geocomposite = $\dfrac{\left(\dfrac{5.4}{13}\right)}{10}$ = 0.04
  (see Chapter 4 for explanation)

Total air flow through geocomposite given by Darcy's law is

$$Q_v = \left[\frac{K_i \gamma A i}{\mu}\right]$$

$$Q_v = \left[\frac{1.2 \times 10^{-5} \times 13 \times 0.025 \times 0.04}{1.75 \times 10^{-5}}\right] = 8.9 \times 10^{-3} \text{ m}^3/\text{s/m width}$$

$(8.9 \text{ l/s/m width})$

There will be friction losses within the system but at the air flow rates that occur within passive systems these are unlikely to be significant and the most convenient way of dealing with these is to apply a factor of safety based on the constrictions within the system.

Passive venting systems can also be designed to varying reference wind speeds (U_r) in the range 0.3–3 m/s and the results compared to the criteria provided in the Partners in Technology report (see Figure 8.4, colour section). In the calculation example provided previously, the equilibrium design gas concentration was 1% v/v and this was achieved at a reference wind speed U_r of 0.55 m/s. This places the system in the good performance category.

An important part of any underfloor ventilation assessment is to ensure that the potential for areas where stagnant air flow conditions can occur is minimised. This means that vents etc. must be carefully positioned and distributed to try to ensure maximum air flow to all locations in the void.

8.6.5 Basement car parks

Basement car parks provide a significant amount of ventilation and this also makes them very effective in protecting against ground gas (see Box 8.7).

Box 8.7 Using a basement car park as gas protection

Car parks should be ventilated passively (or naturally) wherever possible (Institution of Structural Engineers, 2002). The Institution of Structural Engineers (2002) and the Building Regulations Part F require the following ventilation:

1. Provision of vents in the two opposing longer sides with a ventilation area equal to 2.5% of the parking area (on each side and at each level in the car park).

Or

2. Mechanical ventilation to dilute carbon monoxide concentrations to 50 ppm (usually by providing a minimum ventilation of 3–8 air changes per hour depending on usage patterns).

Assume a basement car park that is 30 m × 20 m in plan and 2.5 m high is to be built on a site with a surface emission rate of methane of 0.023 l/h/m²

Volume of car park = 30 m × 20 m × 2.5 m = 1500 m³

Assume it has mechanical ventilation to provide four air changes per hour

Therefore fresh air flow rate, $Q = 1500\,m^3 \times 4 = 6000\,m^3/h$

Surface emission rate (convert to m³) $= \dfrac{0.023}{1000} = 2.3 \times 10^{-5} m^3/h/m^2$

Total surface emission rate of gas into void, $q = 30\,m \times 20\,m \times 2.3 \times 10^{-5} m^3/h/m^2 = 0.0138\,m^3/h$

Required design concentration is 0.25% v/v of methane in the car park.

Design fresh air flow rate required, $Q = 0.0138\,m^3/h \times \left[\dfrac{(100 - 0.25)}{0.25} \right] = 0.0138$
$\times\, 399 = 5.5\,m^3/h$.

The design of the car park gives 6000m³/h which is much greater than the 5.5 m³/h required and therefore the car park will provide a high level of protection against ground gas migration into the building.

8.6.6 Active ventilation sizing

Active ventilation design is simply a matter of providing sufficient fans to give the required air flow for dilution of the ground gas to the required levels. Friction losses in an active system can be higher than passive systems, depending on the air flow rate and constrictions. It may be necessary to estimate the friction losses using the conventional methods that are routinely used in the air conditioning industry.

Box 8.8 Active venting design

Building is $30\,m \times 60\,m$
Design gas concentration $= 76\%$ and design borehole flow rate $= 20\,l/h$
Surface emission rate $= 0.76 \times \dfrac{20}{10} = 1.52\,l/h = 1.52 \times 10^{-3}\,m^3/h$
Design equilibrium concentration in void $= 1\%$
Fresh air flow $Q = 30\,m \times 60\,m \times 1.52 \times 10^{-3}\,m^3/h \left[\dfrac{100 - 1}{1} \right] = 271\,m^3/h$
Place fans at 10 m centres to give evenly distributed airflow so this gives six fans
Each fan must provide an air flow of $\dfrac{271}{6} = \dfrac{45\,m^3}{h}$

It is possible to design passive systems so that the pipework and electrical connections are in place to allow an upgrade to an active system if that is required. The underfloor void of the passive system is monitored. If monitoring indicates that the passive system is not working adequately, active fans can be easily and quickly installed. This has been done on projects where the desk study has identified a potentially high generation potential source but there has been insufficient gas monitoring data for a robust design. It has proved to be a cost-effective way of managing the risk.

Active systems can also be used on lower gas regime sites to enhance the air flow through void formers below large buildings and avoid the need for high-level stacks. However, the increased risk implications associated with a lack of maintenance should be carefully assessed and all active systems should be able to act passively in the event of fan failure.

8.6.7 Positive underfloor pressurisation

The design of positive underfloor pressurisation should provide an estimate of the air flow through the void that is sufficient to maintain the required pressure (see Box 8.9).

The calculations should be used for initial sizing of the fans. Once installed the pressure build up in the void or granular layer (at critical locations chosen by the regulator, not at points where compliance is most likely) should be measured and the air flow adjusted accordingly.

Box 8.9 Positive air flow calculations

An example layout is shown.

Assume type 1 subbase is being used as the air blanket medium (0.15 m thick).

Assume positive pressurisation is required to resist an in ground gas pressure of 10 Pa with a factor of safety of 2. So design gas pressure = 20 Pa. Say pressure at diffuser is 350 kPa, so pressure drop over 10 m to edge of building is 330 kPa.

The layout of the system uses diffuser boxes to direct the air flow to the required points and subfloor monitoring points (probes) to measure the gas concentrations and pressure below the building.

K_i = intrinsic permeability of subbase in m^2 = 2.7×10^{-8} m^2 (from Partners in Technology Report, DETR 1997)

γ = unit weight of air in N/m^3 = 13

μ = viscosity of air in Ns/m^2 = 1.75×10^{-5}

A = surface area over which flow occurs in m^2 = 0.15×6 m = 0.9 m^2 (assume 6 m wide path to edge of building)

$$i = \text{pressure gradient across subbase} = \frac{\left(\frac{330}{13}\right)}{10} = 2.5$$

Total air flow through required to maintain required pressure $Q_v = \left[\frac{K_i \gamma A i}{\mu}\right]$

$$Q_v = \left[\frac{2.7 \times 10^{-8} \times 13 \times 2.5 \times 0.9}{1.75 \times 10^{-5}}\right] \times 3600 = 162 \, \text{m}^3/\text{h flow to each diffuser box}$$

For four boxes this gives $162 \times 4 = 650 \, \text{m}^3/\text{h}$ fan capacity.

It should be noted that the presence of more permeable material will significantly increase the required air flow (e.g. more permeable ground or gravel filled sewer trenches). **In order to maintain a positive pressure at reasonable flow rates a lower permeability material is beneficial. It is unlikely that this will vent passively (passive venting requires material with a high permeability).**

SUMMARY: Design of protection measures

The scope of a gas protection system is based on the results of the risk assessment. A combination of measures is usually used to minimise the risks associated with ground gas. This is so that if one element of the protection fails the others will continue to protect the development. Once the scope has been defined, detailed design of individual elements such as the venting or membranes can be completed. The latest CIRIA guidance has two methods of defining the scope of gas protection: one that is for low-rise housing and a ventilated underfloor void (minimum 150 mm high); the other is for any other type of development.

The system for low-rise housing was developed for the NHBC and the GSV is used to determine the colour-coded classification of a site and the corresponding scope of protection required. This method assumes that all developments have a ventilated underfloor floor void as the minimum protection for the green classification. For all other types of development the scope of gas protection measures is determined by using a modified version of the approach described by Wilson and Card (1999).

If the gas source is adjacent to the development site, the design philosophy should be based primarily on keeping gas out of the site using in ground barriers, and then protecting the buildings. If that is not possible (if the source is below the site) then the main gas protection should be below the building. Passive ventilation to give a clean zone of air around and below a building is the preferred primary method of protection with secondary protection provided by a gas membrane. Active systems should be designed to act passively wherever possible.

The requirement for redundancy in gas protection means that generally at least two different methods are required to ensure protection is maintained if one method fails (e.g. a membrane and ventilated void).

Claims that any single method of can act as two types of protection are untrue. A positive pressurisation system does not provide two levels of protection (a barrier and a ventilation system). If it stops working a second level of protection, such as a membrane, is required.

For larger buildings the use of passive venting becomes more difficult and active venting or pressurisation may be required.

Internal monitoring should be considered a last resort, not a means of protection. The best place for monitoring points is in the underfloor void or vent system.

There is often concern about the durability of membranes at sites where hydrocarbon contamination is present. However, the membrane is not usually in contact with high concentrations of contamination so there is not usually a great risk of damage to membranes.

All ventilation and pressurisation systems should be designed to provide adequate air flow to either dilute gases to the design levels or maintain the necessary pressure below a slab. This chapter has provided guidance on how to complete these calculations.

CHAPTER NINE

Construction and validation

Validation of the installation of gas protection measures is arguably the most important aspect of the design and construction process. This should form part of an overall verification process that includes a design report.

9.1 Design report

The first stage of the QA programme should be the design report. All gas protection designs should be accompanied by a design report that sets out the justification for the parameters used, demonstrates the conceptual model and includes the risk assessment basis on which the protection measures have been designed. The report should include a checklist such as the one in Appendix B to confirm that all the necessary aspects have been covered and the data provided. This will make it easier for regulators to quickly assess schemes.

The design report should also be provided to site managers so that they know why the protection measures are being installed and what is required.

9.2 Common problems with installation

Experience has shown that it is very easy for gas protection measures to be installed incorrectly or for them to be damaged after installation by follow on trades (plumbers, electricians etc.). Independent inspection and QA procedures are therefore vital when installing gas protection. Common problems in the installation of protection in housing have discussed in BRE Report 414 (Johnson, 1999) and in the latest NHBC guidance on ground gas (Boyle and Witherington, 2007).

Common problems include:

- Poor preparation of laying surface for membrane with gaps and projections (Figure 9.1a, colour section)

- Poor sealing around services
- Services located too close to walls making sealing difficult (Figure 9.1b, colour section)
- Air bricks missing from underfloor venting
- Cross-venting omitted in sleeper walls
- Debris from construction blocking vents

Testing of membranes for leaks can increase confidence in the installation but independent installation should still be required. Experience has shown that even quality assured and tested membranes can suffer damage after installation and approval. Examples of damage to membranes or poor installation that has been observed are summarised below:

- Differential settlement occurred between a ground-bearing slab and walls on piled foundations. The membrane laid across the slab and through the walls was torn apart. This can be avoided by using suspended floor slabs or designing the slab to limit the amount of differential movement between it and the building structure
- Difficult detailing around complex structural forms. Careful design and the use of quality assured installers can overcome this
- Follow on trades nailing fixings through membranes (Figure 9.2, colour section)
- Rebar and scaffolding thrown onto membranes that have been laid, prior to casting slab (Figure 9.3, colour section)
- Bricklayers using trowels to retrieve mortar from a cavity have ripped membranes

Follow on trades often cause damage to exposed membranes. The membranes should therefore be protected after laying (there are proprietary boards or geotextile fleeces available for this purpose) or the membrane should be inspected immediately before it is covered over, especially if it is in a housing development where plumbers, plasterboarders etc. have been working over the membrane.

As a final illustration of why validation is so important some quotes from engineers that undertake this kind of work are provided in Box 9.1.

Box 9.1 Quotes from consulting engineers who validate membrane installation

'I'm pretty sure I have photos showing a gas membrane where the seams were sealed with nails, I tell no lie!'

'You could get the photos for the first inspection I did at Site X. I was demonstrating that that service seals left something to be desired ... when I lost my pencil through the hole.'

'There was also that time when the local authority came to inspect one of my gas membrane validation visits ... the gas membrane was blowing across the slab and the contractor had forgotten to tape seal the sheets of plastic.'

Communication and education of site staff is also an important role for consultants. The membrane shown in Figure 9.4 (see colour section) was installed by a general groundworker. However, he was shown how to install the membrane by the consultant who also pointed out all the areas that would be inspected after installation and what would cause it to be rejected.

Conversely the membrane in Figure 9.5 (see colour section) was also installed by a groundworker and shows the problems that can occur when those constructing gas protection do not understand how to do it (poor sealing using incorrect (general adhesive) tape, laid so that it can easily tear when weight is applied to it, incorrect detailing over cavity).

9.3 Validation

Validation of the gas protection works is probably the most important part of any gas protection works. It will be required to avoid the problems discussed in the previous sections and to ensure that the works meet the specification and the risk management objectives. The validation should be carried out by an independent consultant and should be in addition to any QA undertaken by the installation contractor.

Important items that require validation are:

- Membrane installations
- Floor slab construction
- Underfloor venting
- In ground gas venting and barriers

A specific validation plan should be prepared for each site, based on the site specific design and risk assessment. Details of specific problems that often occur and should form part of any validation plan have been provided in Section 9.2.

There are usually changes to the design as a result of circumstances that occur or come to light during construction. Validation helps to manage these and make it clear to others why the changes were made and that any variations do not adversely impact on meeting the risk management objectives. Post-development verification is also a requirement of CLR 11 (Environment Agency, 2004d).

There is very good guidance about validation or verification of remedial works and gas protection measures that has been published by the NHBC and Environment Agency (Boyle and Witherington, 2007; Environment Agency, 2000). A verification report plan checklist, based on guidance from the Environment Agency and NHBC, is provided in Apppendix C.

The key requirements from the NHBC guidance are that for the Amber 2 classification (see Chapter 8) the gas resistant membrane should be:

> installed by certified professionals who should carry out appropriate integrity testing. The most effective post-installation test method is to pressurise the underside of the membrane with an appropriate tracer gas and then sweep the top surface with a suitable gas-detection device (BRE 414). The advantage of this method is that the whole membrane including joints are tested. It has been demonstrated that this test method can detect even very small gas migration routes in the membrane. Any leaks found in the membrane or the joints should be sealed before construction continues.

Membranes with welded joints can also have the joints air tested (see Chapters 7 and 8).

A good guide showing what to look for when inspecting membrane installations has been prepared by Cooper Associates and Smith Grant for Liverpool City Council. Examples of poor installations have been photographed and a verification form and guide has been prepared. The verification form and guide is included in Appendix C. Photographs of poor installation are shown in Appendix D.

Verification of membranes can also include taking samples of materials delivered to site to ensure that they comply with the specification: this is to be encouraged.

9.3.1 Post-construction monitoring

Post-construction monitoring of underfloor voids is usually resisted by housing developers and is not widely undertaken. However, it is possible to monitor the underfloor voids between membrane installation and selling of a property and this has been carried out on a number of sites. Post-construction monitoring has also been used to demonstrate the effectiveness of underfloor voids in reducing the risk of gas ingress where gas protection membranes have been omitted by mistake or where validation has not been undertaken.

Post-construction monitoring is less of an issue with commercial developments and it is widely used to demonstrate the effectiveness of in ground gas migration barriers.

Points to be aware of in post-construction monitoring are:

- If monitoring is required to demonstrate the effectiveness of a subfloor venting or pressurisation system then in ground monitoring in boreholes is also preferred (without boreholes it is difficult to assess if the absence of gas is due to the venting/ pressurisation or due to the absence of ground gas emissions).

- Monitoring in subfloor voids should preferably be carried out via specially installed monitoring points. In open voids these can be retrofitted by drilling through walls and installing small diameter tubes to the centre of the void.
- Pumping during sampling should be of sufficient duration to ensure that a representative sample of the air in the void is obtained (this will depend on the length of the sampling line into the void).
- When installing gas wells to monitor the effectiveness of in ground barriers the wells must only intercept the gas migration pathway and not any other sources of gas that the barrier is not designed to deal with (e.g. made ground).
- Air flow through vents can be measured in some cases (where unobstructed access into the vent is possible by removing any front covers) using an anemometer. This is not normally possible where air brick vents are used.

SUMMARY: Construction and validation

Validation of the installation of gas protection measures is arguably the most important aspect of the design and construction process. This should form part of an overall verification process that includes a design report.

The process is as follows:

(1) Prepare a design report that sets out the justification for the parameters used, demonstrates the conceptual model and includes the risk assessment basis on which the protection measures have been designed
(2) Prepare a validation plan
(3) Validate the installation including gas-resistant membranes and underfloor venting systems. This should be carried out by an independent consultant

It is very easy for gas protection measures to be installed incorrectly or for them to be damaged after installation by follow on trades (e.g. plumbers, electricians etc.). Therefore independent inspection and QA procedures are vital when installing gas protection. Some of the common problems are discussed in this section.

Damaged or leaking membranes can occur due to a variety of causes including differential settlement, complex detailing, follow on trades and bricklayers. Exposed membranes should be protected after laying (there are proprietary boards or geotextile fleeces available for this purpose) or the membrane should be inspected immediately before it is covered over.

Communication and education of site staff is also an important role for consultants. They should explain to site staff exactly what they will be looking for when inspecting membranes and what will cause them to fail the inspection.

Important items that require validation include: membrane installations, floor slab construction, underfloor venting and in ground gas venting and barriers. A specific validation plan should be prepared for each site, based on the site-specific design and risk assessment.

Guidance is provided on what to look for when inspecting membrane installations and underfloor venting as well as what information to include in a verification report. Guidance on post-construction monitoring is also provided.

CHAPTER TEN

Maintenance

All gas protection measures require maintenance. Even passive systems require regular inspections to ensure that vents are not blocked and are still working. The importance of maintenance is highlighted by the example in Chapter 1 where a vent trench became blocked, thus causing an explosion in a building.

The most common maintenance requirements are:

- Inspection of vents to ensure that they are not blocked
- Inspection of vent trenches to ensure that the vents are not blocked
- For gravel-filled trenches with an open surface the gravel will eventually block with debris and will require replacement with clean material
- For vent trenches or other systems with ground-level vent boxes the vents will require regular cleaning out of blockages (typically vegetation or gravel becomes trapped in ground level vents)

All active systems (ventilation and pressurisation) and monitoring and alarm systems are more reliant on maintenance and will require regular inspections and servicing by specialist contractors. Typically this will be every six months and the contractors will check the operation of the whole system and replace or repair items such as sampling lines, monitors, fans etc. as appropriate.

Unfortunately there are numerous examples where active systems have not been maintained after the first few years of operation. This usually occurs as a result of changes in the owner or occupier's staff. The new staff are not aware of the need for the maintenance with the result that it is not carried out. This is why active systems should be designed to vent passively wherever possible.

All systems should be designed to allow for a lack of maintenance and blocking of vents etc. This can be achieved by making an allowance in design calculations for blocked vents. This is more critical for individual housing where the risk of blockage is greater and a more conservative approach should be taken.

Appendix A

A1 Gas monitoring protocol

It is good practice to have a standard measuring protocol to remove as many errors in readings as a result of operator factors as possible. The protocol can be adapted to suit specific circumstances but a record should be made of any variations and they should be applied consistently to any one site. The CIRIA reports give good guidance on a monitoring protocol and this appears to be followed by most leading practitioners in the field. A summary of the data that should be recorded is provided below.

A1.1 Pre-start checks

Record type and serial no of gas monitor, make sure it is in calibration and record date of last factory calibration and date of next due calibration. Record date, time and person undertaking monitoring.

Carry out instrument field calibration and record results. Check general condition of instrument and connecting tubes, filters etc. and record on sheet.

Measure atmospheric pressure, if it is rising or falling and record weather conditions. Record condition of ground (e.g. frozen, saturated, dry etc.).

A1.2 At each sampling point

Record date, time, sample point type (e.g. 19 mm diameter well in window sample hole or 50 mm diameter well in borehole), location or reference no and condition of point (e.g. open, closed, vandalised), depth of well (note: do not measure depth as this will allow gas to escape, take the depth from records).

Record any activities observed in or around site that may affect readings (e.g. excavations, piling etc.)

Monitor borehole flow and pressure first if possible (it may not be possible with some flow meters). Attach monitor to gas tap and then open gas tap and record maximum flow and pressure and then any reduction that occurs and over what time. Record flow rate and pressure for a minimum of one minute and record maximum, minimum and average readings.

During flow monitoring keep monitor shielded from wind as far as possible. Record if wind is affecting readings.

Pump and monitor for methane, carbon dioxide, oxygen and any other gases that are required on the site (e.g. hydrogen sulphide or carbon monoxide). Record maximum and any variations (for example if gas concentrations are rising record the value every 30 s or 1 min. Record until readings stabilise to constant value or for three full volume changes if possible. The monitoring must be undertaken for a sufficient period of time to take a representative sample. For example a typical gas meter will remove 300 cc per min (0.3 l/min) when pumping. Depending on the diameter and length of tubing connecting to a well and the air space volume different minimum times may be appropriate (Table A1).

Table A1 *Illustrative minimum pumping times*

Scenario	Volume (l)	Minimum pumping time	
Connection tube to gas tap 5 mm internal diameter, 2 m long	0.04	8 s	Minimum time to displace air within tube
Well 50 mm diameter, 2 m deep air space	3.9	Until readings stabilise or 39 min	Minimum time to take a sample that comprises three full changes of the well volume (CIRIA Report 131)
Well 150 mm diameter, 4 m deep air space	70.6	Until readings stabilise or 706 min	Minimum time to take a sample that comprises three full changes of the well volume (CIRIA Report 131)

This shows that simply pumping for a set time may not obtain a representative sample.

Table A1 shows that for a 150 mm diameter well with a 4 m deep air space the time to obtain a truly representative sample, based on three air changes, would be 706 min (i.e. nearly 12 h). In such a situation a compromise will normally have to be reached that balances the time to monitor against the benefit gained. If the readings do not stabilise quickly it may be that worse case gas conditions can be predicted from a shorter period of monitoring than the full 706 min. A reasonable maximum time is in the range 5–10 min.

A1.3 Carry out field calibration and record results

CIRIA Report 131 (Crowhurst and Manchester, 1993) recommends measuring the water-table depth before taking gas readings. In the present authors' opinion this should be ignored and gas and flow readings taken before removing a well cap to take water-level readings. This is because the gas regime within the well will be disturbed by removing the cap. In particular, any pressure that has built up within the well will be released.

The most common gas detection instruments in use today are infrared meters. They can measure concentrations of methane and carbon dioxide at the same time and without using up the sample. They are not affected by low oxygen concentrations and also give the most reliable readings of carbon dioxide (Crowhurst and Manchester, 1993). They can, however, be affected by cross-sensitivity if other hydrocarbon vapours or gases are present and in these situations may give misleading results for methane. Whatever type of instrument is used it must be suitable to monitor the target gas in the conditions to be expected and give reliable readings to the required sensitivity and accuracy. Typical requirements are given in Table A2.

Table A2 *Monitoring instrument sensitivity and accuracy*

	Range of sensitivity	Typical accuracy
Methane	0.1–100%	±0.2% @5%, ±1.0% @30% and ±3.0% @100%
LEL	0–100% LEL	±4% LEL
Carbon dioxide	0.1–100%	±0.1% @10%, ±3.0% @50% and ±3% @100%
Hydrogen sulphide	0–200 ppm	5% of full scale
Oxygen	0.1–25%	±0.5%
Atmospheric pressure	800–1200 mb	±5 mbar
Borehole flow rate	+30 to −10 l/h	0.1 l/h resolution
Borehole pressure	+300 to −30 Pa	

Appendix B

Design report checklist (based on Environment Agency guidance on requirements for land contamination reports, Environment Agency 2005c)

Checklist to ensure that all relevant information has been provided in this report.

Design reporting requirements:

Contents:	Provided?
Report objectives	Yes ☐/No ☐
Site location map and National Grid Reference	Yes ☐/No ☐
Site layout plans	Yes ☐/No ☐
Site area in hectares	Yes ☐/No ☐
Description of site and surroundings	Yes ☐/No ☐
Details of desk study researches undertaken	Yes ☐/No ☐
Information on past and current activities at the site	Yes ☐/No ☐
Details of intended future use of the site	Yes ☐/No ☐
Unique references for all relevant planning applications or permissions at the site	Yes ☐/No ☐
Historical Ordnance Survey maps and site plans and if available, aerial photographs	Yes ☐/No ☐
Environmental setting including:	
• Superficial deposits and solid geology	Yes ☐/No ☐
• Hydrology	Yes ☐/No ☐
• Hydrogeology (including the interaction between all relevant shallow and deep groundwaters and how they flow to potential receptors)	Yes ☐/No ☐
• Location and status of relevant receptors	Yes ☐/No ☐

Contents:	Provided?
Information on site drainage and other man-made potential pollutant pathways, e.g. underground services	Yes ☐/No ☐
Identification of potential gas or vapour source areas	Yes ☐/No ☐
Consultations with the local authority	Yes ☐/No ☐
Consultations with the Environment Agency	Yes ☐/No ☐
Consultations with other appropriate bodies	Yes ☐/No ☐
Review and summary of previous reports, with report references	Yes ☐/No ☐
Outline conceptual model with nature and location of gas sources, migration pathways and receptors clearly identified	Yes ☐/No ☐
Description of possible pollutant linkages for ground gas	Yes ☐/No ☐
Identification of potentially unacceptable risks posed by ground gas, including criteria used to identify those risks	Yes ☐/No ☐
Discussion of uncertainties and gaps in information	Yes ☐/No ☐
Description and justification of next steps proposed at the site, e.g. carry out site investigation and quantitative risk assessment	Yes ☐/No ☐

Notes

All plans and historical maps extracts must be large scale, to scale, with a north point, and clearly show the site boundary

The report should be prepared by a suitably qualified professional and should contain evidence of their credentials

Appendix C

Verification report checklist (based on Environment Agency guidance on requirements for land contamination reports (Environment Agency, 2005c)

Checklist to ensure that all relevant inspections have been carried out and all relevant information has been provided in this report.

Verification reporting requirements:

Contents:	Provided?
Report objectives	Yes ☐/No ☐
Site location map and National Grid reference	Yes ☐/No ☐
Site layout plans	Yes ☐/No ☐
Review and summary of previous reports, with references	Yes ☐/No ☐
Scope of remediation works to be undertaken and any design details required to inform the verification plan	Yes ☐/No ☐
Description of what constitutes completion for the remedial works and how completion will be verified (e.g. inspections, photographs of completed membranes and air bricks etc.)	Yes ☐/No ☐
Data gathering requirements to demonstrate that site remediation criteria are achieved for each relevant pollutant linkage, such as: • Inspection and monitoring strategy, including: frequency and timing of visits and information to be recorded (e.g. photographs); post-completion verification gas monitoring requirements and frequency (e.g. in underfloor voids or in the ground behind a gas barrier) • How site observations will be recorded • Testing and monitoring quality assurance and control requirements	Yes ☐/No ☐
Performance testing required, e.g. for membranes or void formers	Yes ☐/No ☐

Contents:	Provided?
Plans showing proposed sampling and monitoring point points where post-installation monitoring is proposed	Yes □/No □
Explanation of how compliance with regulatory consents has been achieved	Yes □/No □
Actions taken where: • Test results and monitoring data show that the remediation activities have not achieved the remediation criteria derived for relevant pollutant linkages • Site works vary from those anticipated in the design report or implementation plan	Yes □/No □

All plans and historical maps extracts must be large scale, to scale, with a north point, and clearly show the site boundary

The report should be prepared by a suitably qualified independent professional and should contain evidence of their credentials. Consultants undertaking independent validation should be completely independent (for example local authorities should ask for confirmation that consultants do not have any business interests in the installers or vice versa)

Points for inspection of gas protection measures (gas-resistant membrane on a cast in situ reinforced concrete suspended floor slab (provided by Cooper Associates (Nik Reynolds and Ben Hill), prepared for Liverpool City Council)

1) Check general condition of membrane for punctures/tears etc. Take panoramic photograph showing general condition of the membrane

2) Confirm membrane product type and the membrane is sufficient for methane resistance. Photograph both sides of gas membrane

3) Check joints between rolls of membrane to ensure 150 mm overlap (or as recommended by manufacturer), use of correct double sided tape approved by manufacturer (typically butyl tape). This may not be visible and may have to be determined by feel

4) Confirm presence of swan neck vents on sides and rear of property (NB: in this case front of property did not require venting)

5) In this case it was necessary to lift the membrane to confirm the presence of sand blinding over the sacrificial 1200 gauge (300 micron) DPM. Lift DPM to confirm granular void forming the stone venting layer

6) Check top hat is sealed to membrane and pipework. If the pre-formed top hat is not tight to the pipework then also confirm presence of jubilee clip beneath butyl tape

7) Inspect product labels to confirm manufacturers and product types conform to design and specification. Take samples for independent testing where appropriate

8) If product labels are absent then inspect stores for materials and ask for delivery notes

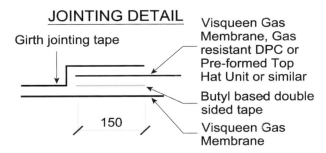

JOINTING DETAIL

Girth jointing tape

Visqueen Gas Membrane, Gas resistant DPC or Pre-formed Top Hat Unit or similar

Butyl based double sided tape

Visqueen Gas Membrane

150

It should be noted that this specification relates to ground gas mitigation measures without a fully vented below-floor slab void. This inspection guide refers to suspended floor slab mitigation measures. Prior to undertaking any ground floor slab works, all designs for the membrane type, form of mitigation, the robust typical details (cross-section diagram) for the proposed works, including all service entry points will require approval by the local authority.

In the example above, Visqueen has been utilised as the ground gas membrane. There are other products which are as suitable for use (see Chapter 7).

Similar inspection protocols can be developed for different types of floor slab construction and gas protection measures.

LIVERPOOL CITY COUNCIL
PUBLIC PROTECTION DIVISION

Gas Protection Validation Record

coopers

One record sheet to be completed for each plot inspected

job number		design source/ref:		specification source/ref:		other documents attached	✓ / ✗ ☐
site name / location		building use:	residential	commercial	other (describe)		
plot number/s		building description:	no. of storeys =	detached	semi-detached ☐	terrace ☐	apartment block ☐
compiled by:		gas protection type:	active / passive	foundation type:	☐	suspended floor / raft / other	

Ventilated sub-floor (if present)	✓ / ✗	inspection date/time:		inspected by:		photographed:	✓ / ✗ ☐
		Notes/recommendations (see guide below)					
void former type	☐	1.					
height of void space	☐	2.					
gravel type	☐	3.					
pipe size and spacing	☐	4.					
external wall airbricks	☐	5.					
internal sleeper walls	☐	6.					
external vent trenches / ducts	☐	7.					

Gas barrier	✓ / ✗	inspection date/time:		inspected by:		photographed:	✓ / ✗ ☐
		Notes/recommendations					
membrane type	☐	8.					
extent of coverage	☐	9.					
underside of membrane	☐	10.					
slab/membrane condition	☐	11.					
laps and joints	☐	12.					

LIVERPOOL CITY COUNCIL
PUBLIC PROTECTION DIVISION   ☰coopers

Gas Protection Validation Record

damp-proof course	13.
service entries and seals	14.
cavity Inspection	15.

Guide notes:
(Also refer to pdf guidance sheets and specifications appended to record sheet)

1	void former type	proprietary type - manufacturer and specification, in accordance with design?, installed properly without damage?
2	height of void space	height of proprietary former or constructed ventilation space below suspended floor - note any debris on void / obstructions to air flow, note formation surface soil type (e.g. crushed concrete/brick), any evidence of flooding
3	gravel type	gravel type, if used (limestone / granite etc.) and any specification (e.g. 6F2), typical particle dimensions (mm), apparent fines content (low/high), compaction (loose/dense), waterlogging / contamination by clay, organic matter, other debris. Take photographs of stockpile close up shot of stone with tape measure. Alternatively check details on delivery tickets for stone. Take photographs of adjacent plots if at this stage of construction. Check depth of stone conforms to at least 300mm if visible.
4	pipe size and spacing	diameter in mm; material type (e.g. PVC); slotted or perforated; positioning and spacing / separation and jointing as on design drawing - if not sketch arrangement - do pipes connect with external (telescopic / swan-neck) vents? Take photographs of vents on external walls for each plot. (May be possible to photograph other plots on site which are at stage of installing vents. Will be useful for these plots later on).
5	external wall airbricks	check numbers, size and positions as design drawing (if not shown, make sketch; check for blockage, e.g. by mortar, or soil / pavings etc.
6	internal sleeper walls	check for ventilation holes - e.g. honeycombe brickwork or pipe crossings - note size, spacing and location - in accordance with design?
7	external vent trenches / ducts	check whether located and constructed in accordance with design drawings; if open-topped gravel, note gravel type and presence of fines / contamination; if pipe or other vents - check positions and construction for functionality and absence of blockages - vents may be built over
8	membrane type	note manufacturer and product specification, including batch / roll numbers if present - in accordance with specification? Check stock storage arrangements - protected from dirt and damage?
9	extent of coverage	if membrane is incomplete, further inspection will be required - note areas completed / incomplete - is membrane fully visible or have internal walls been constructed over membrane / screed placed?
10	underside of membrane	Where necessary, for example, when using a granular blanket as a ventilation layer, check the underside of the membrane has adequate protection e.g. minimum 50mm no fines concrete blinding layer or appropriate geo-textile (see also below)

LIVERPOOL CITY COUNCIL
PUBLIC PROTECTION DIVISION
coopers

Gas Protection Validation Record

11	slab/membrane condition	record presence of debris / rough surfaces, in particular sharp projections, below or above membrane; record locations of all punctures or repairs, note arrangements to protect membrane surface from traffic / tools and equipment / materials, and temporary weighting down of membrane, e.g. use of boards - record evidence of footprints / tracks on membrane surface, creases or water/wind damage. Take photographs of each plot inspected.
12	laps and joints	check the all joints are lapped and sealed in accordance with manufacturer's requirements / specification, particularly where creases/folds are present (usually minimum 150mm laps with double-sided tape between sheets, and single-sided tape on top surface; note size of sheets and frequency of edge seals). Take photographs of jointing for each plot.
13	damp-proof course	record DPC manufacturer and product code - usually integrated with the membrane; measure the DPC projection from external wall in mm; check laps and seals between membrane and DPC - note any potential stress points and tension between the two; check for damage to DPC
14	service entries and seals	note number, position and diameter of service entries - check top hat seal arrangements in accordance with design / specification (laps and seals between top hat and floor membrane, pipe upstand is usually a minimum 150mm) check , with jubilee clips to secure top hat seal to pipe - note presence of clips and tightness of connections. Take photographs for all plots inspected.
15	cavity Inspection	check gas membrane or gas resistant dpc is taken across cavity. Check for rips across cavity. Check for jointing detail of gas resistant dpc or membrane across cavity to main membrane. Take photographs for all plots inspected.

This plot has PASSED / FAILED* inspection.

(Any proposed remedial works will be noted in the 'Remarks' column on this form).

An addition inspection visit IS / IS NOT* required for this plot.

Site Manager: .... Signed:...........................

(Print Name)

* Delete as appropriate

Appendix D

Photographs of examples of poor installation (provided by Cooper Associates (Nik Reynolds and Ben Hill))

Raft with gas membrane sealed to gas-resistant Damp Proof Course (DPC) spanning walls/cavity

Gas-resistant DPC absent under internal wall

Use of non-gas-resistant DPC (right of picture) compared to gas-resistant DPC (left of picture)

Girth jointing tape absent between sections of membrane (although butyl tape visible)

Tears in gas resistant membrane

Gas membrane not sealed

Gas membrane not sealed and gas-resistant DPC absent

No top hat (formed or pre-formed) around service entry. General adhesive tape used to seal overlaps of membrane. Butyl seal absent on joints. Membrane not sealed to tape

Top hat split, not sealed to service entry or membrane

General adhesive tape failing to adhere to membrane rather than use of butyl tape and girth jointing strap. (Note that the specific tape recommended by the membrane manufacturer must be used. Many tapes will not adhere to HDPE or PP)

Membrane not sealed to gas-resistant DPC

Cast in situ reinforced concrete suspended slab with passive venting comprising gravel venting layer with sacrificial DPM, sand blinding layer and gas-resistant DPM (provided by Cooper Associates (Nik Reynolds and Ben Hill))

Use of non-gas-resistant DPM (blue backing colour) which is same colour as the gas-resistant membrane (silver backing colour). The gas-resistant membrane was used at the edges to cross the cavity so that after casting the concrete it would appear that a gas-resistant membrane had been placed below the whole slab

Separation membrane (to keep sand out of granular venting material) absent with sand blinding filling the voids in the granular material

Top hat split open to fit service entry

Butyl tape used to span over joints of membrane rather than being paced in between sheets

Joints not sealed and membranes
torn, as visible after casting of slab

Seals absent over wall cavity

No top hats sealing service entry

Use of standard butyl tape to seal
service penetration rather than a
pre-formed top hat

Bibliography

AGS (1998) *The AGS code of conduct for site investigation.* Association of Geotechnical and Geo-environmental Specialists, Beckenham, UK

AGS (2000) *Guidelines for combined geoenvironmental and geotechnical investigations.* Association of Geotechnical and Geoenvironmental Specialists, Beckenham, UK.

ASTM (2003) *Standard test method for determining gas permeability characteristics of plastic film and sheeting.* ASTM D1434-82(2003). American Society for Testing and Materials.

ASTM (2004) *Risk-based corrective action.* ASTM Standard E2081-00(2004). American Society for Testing and Materials, Pennsylvania, USA.

Bailer L. C. (1947) Gas fire in sewer manhole traced to sanitary-fill operations. *Engineering News-Record.* December 25, 1947. p 51.

Bannon M. P. and Hooker P. J. (1993) *Methane: its occurrence and hazards in construction.* CIRIA Report 130, CIRIA, London, UK.

Barry D. L., Summersgill I. M., Gregory R. G. and Hellawell E. (2001) *Remedial engineering for closed landfill sites.* CIRIA Report C557, CIRIA, London, UK.

Boltze U. and de Freitas M. H. (1996) Changes in atmospheric pressure associated with dangerous emissions from gas generating disposal sites. The 'explosion risk threshold' concept'. *Proceedings of the Institution of Civil Engineers-Geotechnical Engineering,* **119**, 177–181.

Boyle R. and Witherington P. (2007) *Guidance on evaluation of development proposals in sites where methane and carbon dioxide are present.* Report Edition No 04. March 2007, National House Building Council, Amersham, UK.

BRE (1991) *Construction of new buildings on gas-contaminated land.* BRE Report 212, Building Research Establishment, Watford, UK.

BRE (1999) *Radon: Guidance on protective measures for new dwellings.* BRE Report 211, 3rd edn. Construction Research Communications Limited, BRE Press, Berkshine, UK.

BSI (1990) *Code of practice for protection of structures against water from the ground.* BS8102: 1990. British Standards Institution, London, UK.

BSI (1996) *Physical testing of rubber. Determination of permeability to gases.* BS 903, Part A30: 1996. British Standards Institution, London, UK.

BSI (1999) *Code of practice for site investigations.* BS 5930:1999. British Standards Institution, London, UK.

BSI (2001) *Investigation of potentially contaminated sites – Code of Practice.* BS 10175:2001. British Standards Institution, London, UK.

BSI (2003) *Soil quality – sampling, Part 7. Sampling of soil gas.* BS 10381:2003. British Standards Institution, London, UK.

Card G. B. (1995) *Protecting development from methane.* CIRIA Report 149, CIRIA, London, UK.

Chartwell (2002) Michigan DEQ investigates gas at explosive landfill site. *Chartwell's Weekly News Update,* August 29-September 4, 2002.

Council of the City of Los Angeles (2004) *Los Angeles Municipal Code, Ordinance No. 175790,* Council of the City of Los Angeles, at its meeting of February 12, 2004. Los Angeles, CA, USA.

Crawford J F and Smith P G (1985) *Landfill technology.* Butterworths, London, UK.

Creedy D., Sceal J. and Sizer K. (1996) *Methane and other gases from disused coal mines: the planning response technical report.* Wardell Armstrong, DoE, London, UK.

Crowhurst D. (1987) *Measurement of gases from contaminated land.* Building Research Establishment, Watford, UK.

Crowhurst D. and Manchester S. J. (1993) *The measurement of methane and other gases from the ground.* CIRIA Report 131, CIRIA, London, UK.

DEFRA and Environment Agency (2004) *Model procedures for the management of land contamination.* Contaminated Land Report 11, Environment Agency, Bristol, UK.

Department of Environment (1994) *Waste management paper No 26, landfill completion,* HMSO.

Department of Environment (1989) *Waste management paper No 27, landfill gas.* The Stationery Office, Norwich, UK.

Department of Environment (1994) *Sampling strategies for contaminated land.* Contaminated Land Research Report No 4, Department of Environment, London, UK.

DETR and Partners in Technology (1997) *Passive venting of soil gases beneath buildings, guide for design.* Research report, Volume I, Ove Arup and Partners. London, UK.

DETR (2000) *Environmental Protection Act 1990: Part 2A Contaminated Land.* Circular 02/2000, Department of the Environment, Transport and the Regions, London, UK.

Environment Agency and NHBC (2000) *Guidance for the safe development of housing on land affected by contamination.* R&D Publication 66, Environment Agency, Bristol, UK.

Environment Agency (2001) *Guide to good practice for the development of conceptual models and the selection and application of mathematical models of contaminant transport processes in the sub surface.* NGCL Report NC/99/38/2, Environment Agency, Bristol, UK.

Environment Agency (2002) *GasSim -landfill gas risk assessment tool.* R&D Project P1-295, Environment Agency, Bristol, UK.

Environment Agency (2003a) *Guidance on landfill completion; a consultation.* Environment Agency, Bristol, UK.

Environment Agency (2003b) *Integrated Pollution Prevention Control (IPPC). Environmental assessment and appraisal of BAT.* Horizontal Guidance Note IPPC H1.

Environment Agency (2003c) *Consultation on Agency policy: Building development on or within 250 metres of a landfill site,* Environment Agency, Bristol, UK.

Environment Agency (2004a) *Guidance on the management of landfill gas* . LFTGN 03, Environment Agency, Bristol, UK.

Environment Agency (2004b) *Update on estimating vapour intrusion into buildings.* CLEA Briefing Note 2. Available on-line at http://www.environment-agency. gov.uk/commondata/105385/soil_vapour_intrusion_749183.pdf (last accessed 10 June 2004).

Environment Agency (2004c) *Guidance for monitoring trace components in landfill gas.* Environment Agency, Bristol, UK.

Environment Agency (2004d) *Model procedures for the management of land contamination and associated documentation. CLR 11.* Environment Agency, Bristol, UK.

Environment Agency (2005a) *Review of building parameters for the development of a soil vapour intrusion model.* Environment Agency, Bristol, UK.

Environment Agency (2005b) *Update of supporting values and assumptions describing UK building stock. CLEA briefing note 3.*Available on-line at http://www.environment-agency.gov.uk/commondata/acrobat/bn3_904797.pdf (last accessed 19 January 2005).

Environment Agency (2005c) *Environment Agency guidance on requirements for land contamination reports.* July 2005. Environment Agency, Bristol, UK.

Environment Agency (2005d) *Guidance on monitoring MBT and other pre-treatment processes for the landfill allowances schemes (England and Wales).* August 2005. Environment Agency, Bristol, UK.

Gregory R. G., Revans A. J., Hill M. D., Meadows M. P., Paul L. and Ferguson C. C. (1999) *A framework to assess the risks to human health and the environment from landfill gas.* Environment Agency Technical Report P271, under contract CWM 168/98.

Godson J. A. E. and Witherington P. J. (1996) *Evaluation of risk associated with hazardous ground gases.* Fugro Environmental, Manchester, UK.

Harries C. R., McEntee J. M. and Witherington P. J. (1995) *Interpreting measurements of gas in the ground.* CIRIA Report 151, CIRIA, London

Hartless R. (2004) *Developing a risk assessment framework for landfill gas: calculating the probability of a landfill gas explosion.* Waste 2004, Integrated waste management and pollution control: policy and practice, research and solutions. 28-30 September 2004, Stratford-upon-Avon. UK.

Health and Safety Executive (2003) *Review of landfill gas: Incidents and guidance.* DIN TD5/030, 31 October 2003, Health and Safely Executive, London, UK.

Hesse P. R. (1971) *A textbook of soil chemical analysis.* Murray, London, UK.

Hickman H. L. (2001) A brief history of solid waste management in the US, 1950 to 2000. *MSW Management* Jan/Feb 2001. Santa Barbara, CA, USA.

Hooker P. J. and Bannon M. P. (1993) *Methane: its occurrence and hazards in construction.* CIRIA Report 130, CIRIA, London, UK.

IStructE (2002) *Design recommendations for multi-storey and underground car parks.* 3rd edn. Institution of Structural Engineers, London, UK.

IWM (1998) *The monitoring of landfill gas,* 2nd edn. Institute of Wastes Management, Northampton, UK.

Johnson J. A. (1995) *Water-resisting basements.* CIRIA Report 139. CIRIA, London, UK.

Johnson P. C. and Ettinger R. A. (1991) Heuristic model for predicting the intrusion rate of contaminant vapours in buildings. *Environmental Science and Technology*, **25**, 1445–1452.

Johnson R. (1999) *Protective measures for housing on gas-contaminated land.* BRE Report 414, Building Research Establishment, Watford, UK.

Kitcherside M. A., Glen E. and Webster A. J. F. (2000) FibreCap: an improved method for the rapid analysis of fibre in feeding stuffs. *Animal Feed Science and Technology*, **86**, 125–132.

Lambe T. W. and Whitman R. V. (1979) *Soil mechanics. SI version.* Wiley, New York, USA.

Nastev M. R. (1998) Modelling landfill gas generation and migration in sanitary landfills and geological formations. Ph.D. dissertation, Laval University, Quebec, Canada.

New York Times (1984) Ohio homes condemned by gas bring a fight over compensation. *New York Times* 11 November 1984.

Office of the Deputy Prime Minister (2004a) *Site preparation and resistance to contaminants and moisture, Approved Document C.* The Building Regulations 2000 updated Part C, 2004, The Stationery Office, Norwich, UK.

Office of the Deputy Prime Minister (2004b) *Planning policy statement 23: Planning and pollution control.* The Stationery Office, Norwich, UK.

O'Riordan N. J. and Milloy C. J. (1995) *Risk assessment for methane and other gases from the ground.* CIRIA Report 152, CIRIA, London, UK.

Owen R. and Paul V. (1998) Gas protection measures for buildings, methodology for the quantitative design of gas dispersal layers, Department of the Environment, Transport and the Regions, London, UK.

Pecksen G. N. (1986) *Methane and the development of derelict land.* London Environmental Supplement, Summer 1985, No.13 London Scientific Services, Land Pollution Group, London, UK.

Personal communication (2005) Personal communication from Paul Culleton.

Privett K. D., Williams S. C. and Hodges R. A. (1996) *Barriers, liners and cover systems for containment and control of land contamination.* Special Publication 124, CIRIA, London, UK.

Pueboobpaphan S. and Toshihiko M. (2007) *Assessment of biodegradability of waste in old landfill.* From http://ws3-er.eng.hokudai.ac.jp/egpsee/alumni/abstracts/Suthathip.pdf (last accessed May 2007).

Raybould J. G., Rowan D. L. and Barry D. L. (1995) *Methane investigation strategies.* Report 150, CIRIA, London, UK.

Rudland D. J., Lancefield R. M. and Mayell P. N. (2001) *Contaminated land risk assessment.* CIRIA Report C552, CIRIA, London, UK.

SEPA (2007) *Part 2A of the Environment Protection Act 1990, common questions and answers*. Scottish Environmental Protection Agency, Stirling, UK. www.sepa.org.uk/contaminated-land/partii/3.10.htm (last accessed May 2007).

Sladen J. A., Parker A. and Dorrell G. L. (2001) Quantifying risks due to ground gas on brownfield sites. *Land Contamination and Reclamation*, **9**(2), 191–207.

State of California (2005) *Advisory on methane assessment and common remedies at school sites*. California Department of Toxic Substances Control, School property evaluation and cleanup division. Glendale, CA, USA.

UNFCCC (2005) *Avoided emissions from organic waste through composting*. Revision to the approved baseline methodology AM0025. United Nations Framework Convention on Climate Change, AM 0025/V2, 28 November 2005. UN, New York, USA.

Wheeler S. J. (1988) A conceptual model for soils containing large gas bubbles. *Geotechnique* **38**, 389–397.

Wilson S. A. and Card G. B. (1999) Reliability and risk in gas protection design. *Ground Engineering*, **32**(2), 33–36.

Wilson S. A., Card G. B. and Haines S. (2004) *Gas protection – a common sense approach*. Society of Chemical Industry. Contaminated Land Achievements and Aspirations. 12–15 September 2004, Loughborough, UK.

Wilson S. A. and Haines S. (2005) Site investigation and monitoring for ground gas assessment – back to basics. *Land Contamination and Reclamation*, **13** (3), 211–22.

Wilson S., Oliver S., Mallett H., Hutchings H. and Card G. (2006) *Assessing risks posed by hazardous ground gases to buildings*. CIRIA Report C659. CIRIA, London, UK.

Wilson S., Oliver S., Mallett H., Hutchings H. and Card G. (2007) *Assessing risks posed by hazardous ground gases to buildings*. CIRIA Report C665. (Note this is C659 reissued due to a number of editorial changes, but the technical content is unchanged). CIRIA, London, UK.

www.landfill-gas.com (2007) *The landfill gas explosions which shocked the UK's waste management industry into action* (last accessed May 2007).

Index